Attainment's EXPLORE math

Judi Kinney

Explore Math

By Judi Kinney

Graphic Design by Jo Reynolds

An Attainment Company Publication
©2010 Attainment Company, Inc. All rights reserved.
Printed in the United States of America
ISBN 1-57861-695-6

Attainment Company

P.O. Box 930160 • Verona, Wisconsin 53593-0160 USA
Phone: 800-327-4269 • Fax: 800.942.3865
www.AttainmentCompany.com

Reproducible resources within this material may be photocopied
for personal and educational use.

Attainment's EXPLORE math

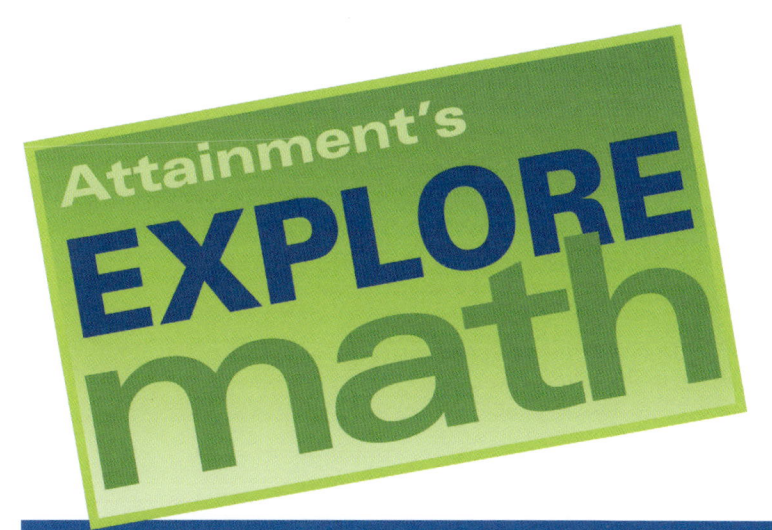

Table of Contents

1 Vocabulary 5

2 0-12. 17

3 0-18. 53

4 0-100 89

5 0-1000 127

6 Fractions 157

Chapter 1
Vocabulary

9, 8, 7, 6 …

Chapter 1 • Vocabulary

Vocabulary 1

addition $$3 \oplus 4 = 7$$	**altogether** $$3 \oplus 2 = 5$$
behind ↓	**between** ↓
counting back $$9, 8, 7, 6 \ldots$$	**counting on** $$\ldots 5, 6, 7, 8$$
difference $$8 \ominus 4 = 4$$	**earlier** (one hour)

Vocabulary 2

equals	first
6 + 5 ⊜ 11 7 − 3 ⊜ 4	

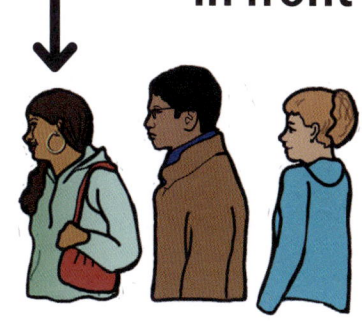

(how many) **in all**	**in front of**
5 ⊕ 2 = 7	

label	last
8 points + 10 points = 18 points	

later	least

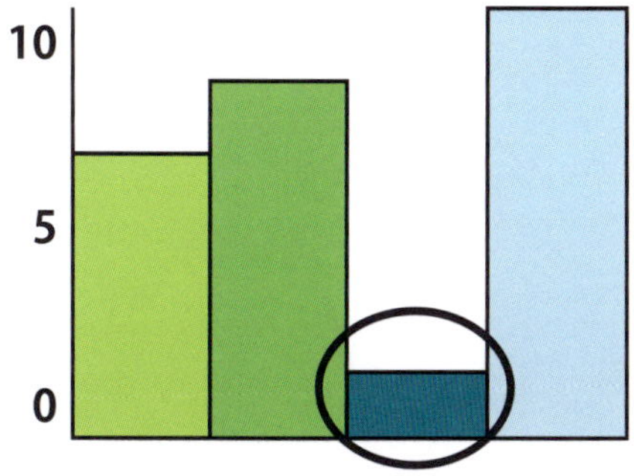

▼ Note: When the words "how many more" appear in the problem, count on or count back.

Chapter 1 • Vocabulary

Vocabulary 3

left 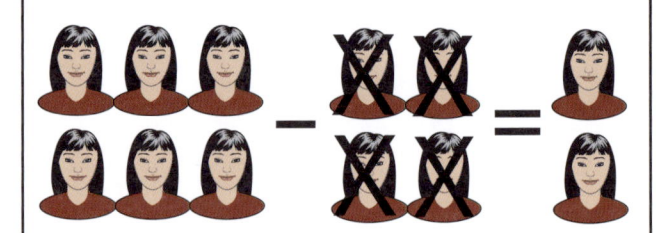	**minus** —
most 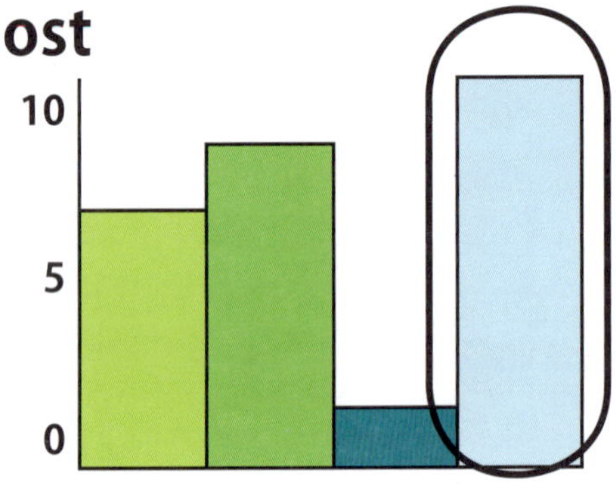	**plus** +
remainder $12 - 9 = ③$	**subtraction** —
sum $14 + 3 = ⑰$	**total amount** 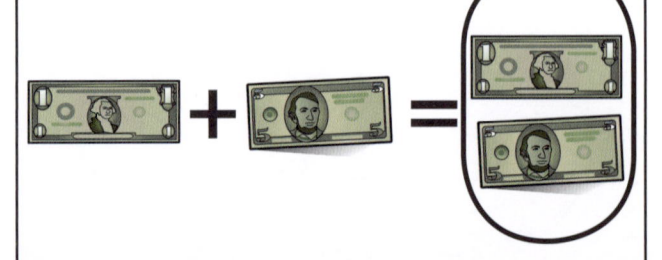

Chapter 1 • Vocabulary

Vocabulary 4

word problem

Kevin has 4 [dollar bill].

He earned 3 more [dollar bill].

How many [dollar bill] did Kevin earn in all?

▼ Note: Use the blank cards for additional words or to make your own cards.

Addition Words +

Date _____

Directions: When reading a word problem, these words tell you to add.

1. altogether

2. how many

3. in all

4. together

5. total

6. sum

7. _____

8. _____

9. _____

10. _____

▼ Note: Add more words if needed, using the blank lines.

Chapter 1 • Vocabulary

Subtraction Words −

Date _____

Directions: When reading a word problem, these words tell you to subtract.

1. difference

2. how many are left

3. left

4. remain

5. remainder

6. _____

7. _____

8. _____

9. _____

10. _____

▼ Note: Add more words if needed, using the blank lines.

Chapter 1 • Vocabulary

Where Am I?

Date _____

Directions: ◯ the number or person that matches the **underlined** word.

1. the **first** number.

   2, 4, 6, 8, 10, 12

2. the number that is in **between.**

   10, 11, 12

3. the **last** number.

   2, 3, 4, 5, 6

4. the person that is **behind.**

5. the person **in front of** the line.

6. the **last** person.

7. the boy that is **between** 2 girls.

Chapter 1 • Vocabulary

Finding Addition Words

Date _____

Directions: Use the word charts or vocabulary word cards and ◯ the addition words. Challenge: Solve.

Problem: **Show work:**

1. **Money in Hand**

   What is the total amount?

2. **Home Town Team**

1st half	2nd half	total score
2	1	?

   What is the total score?

3. **Earning Money**

   Kevin has 4 [$1].

   He earned 3 more [$1].

   How many [$1] did Kevin earn in all?

4. **On Sale!**

   Melanie bought a t-shirt for $6.00.

   She bought a of pair of sunglasses for $5.00.

   How much money did Melanie spend altogether?

   1   2   3   4   5   6   7   8   9   10   11   12

Chapter 1 • Vocabulary

13

Subtraction Words

Date _____

Directions: Use the word chart or vocabulary word cards and ◯ the subtraction words. Challenge: Solve.

Problem:	Show work:
1. Final Softball Score HOME 12 AWAY 9 Find the difference in the two scores.	
2. Fishing Julie had 7 🎣🎣🎣🎣🎣🎣🎣. She gave 5 away. 🎣🎣🎣🎣🎣 (X'd out) How many 🎣 does Julie have left?	
3. Roadside Stand There are 8 🎃🎃🎃🎃🎃🎃🎃🎃. 3 🎃🎃🎃 were sold. How many 🎃 are left?	
4. Basketball Umberto wants to go to the game. He has $6.00. The tickets cost $4.00. How much money does Umberto have left after he buys a ticket?	

1 2 3 4 5 6 7 8 9 10 11 12 13 14 15 16 17 18

Chapter 1 • Vocabulary

Add or Subtract? 1

Date _____

Directions: 1. ◯ the important math words. 2. ◯ add or subtract
3. Challenge: Solve the problem.

Problem: **Show work:**

1. **Garage Sale**

   Tamaka wanted to sell some of her old 🎵.

   She had 12 🎵.

   After the garage sale she had 3 🎵 left. How many did she sell?

   add   subtract

2. **Collecting Old Nickels**

   Michael has 6 🪙.

   He bought 2 more 🪙 at a coin shop.

   How many 🪙 does he have altogether?

   add   subtract

3. **N.F.L. Cards**

   Ethan has 8 🃏.

   His grandmother gave him 2 more 🃏.

   How many 🃏 does Ethan have in all?

   add   subtract

4. **Home Game**

   | HOME | 12 |
   | AWAY | 9 |

   Find the difference between the scores.

   add   subtract

1 2 3 4 5 6 7 8 9 10 11 12 13 14 15 16 17 18

Chapter 1 • Vocabulary

Add or Subtract? 2

Date _____

Directions: 1. ◯ the important math words. 2. ◯ add or subtract
3. Challenge: Solve the problem.

Problem:	Show work:
1. Pick Your Own Harold and Maude picked ▢ boxes of blueberries. Harold picked 4 ▢ boxes of berries. Maude picked 6 ▢ boxes of blueberries. How many boxes did they pick in all? 　　　add　　　subtract	
2. Lunch Count 17 ▢ children ate hot dogs for lunch. 8 ▢ children ate hamburgers. How many more ▢ children ate hot dogs? 　　　add　　　subtract	
3. Favorite Sports 6 ▢ boys like to play football. 4 ▢ boys like to play basketball. How many more ▢ boys like to play football than basketball? 　　　add　　　subtract	
4. N.B.A. Cards Jack has 10 ▢ N.B.A. cards. His aunt gave him 4 more ▢ N.B.A. cards. How many ▢ cards does Ethan have altogether? 　　　add　　　subtract	

1　2　3　4　5　6　7　8　9　10　11　12　13　14　15　16　17　18

16

Chapter 1 • Vocabulary

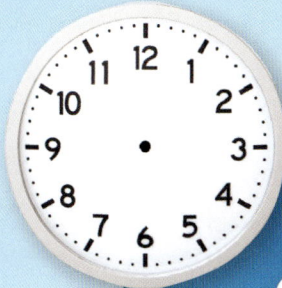

Chapter 2
0 — 12

A Graphic Novel

Date _____

Directions: Nicky bought a graphic novel at a garage sale. ◯ the amount of [1] he needs to pay for the novel.

$7.00

Bonus: ◯ the least amount of money Nicky needs to pay for the novel.

| 1 | 2 | 3 | 4 | 5 | 6 | 7 | 8 | 9 | 10 | 11 | 12 |

Chapter 2 • 0 — 12

Elm Street

Date _____

Directions: Help the mail carrier deliver the mail.
Read the map of Elm Street and answer the questions below.

Hint: You need to count by 2 to find the missing number.

Answer:

1. Write the missing number on the mailbox.
2. ◯ the blue house.
3. Put an **X** on the tallest house.
4. Put a ✔ on the house farthest from the park.

| 1 | 2 | 3 | 4 | 5 | 6 | 7 | 8 | 9 | 10 | 11 | 12 |

Chapter 2 • 0 — 12

Neighborhood Map

Date _____

Directions: Directions: Mr. Jackson delivers mail in the neighborhood. Read the map and answer the questions below.

Answer:

1. Draw a path from the mail truck to the post office.
   How many blocks from the mail truck to the post office? _____ blocks

2. Draw a path from Indira's house to the school.
   How many blocks does she have to walk to school? _____ blocks

3. ◯ the 🏞️ intersection of Oak and First Street.

4. Put an **X** on the 🚦 streetlight.

5. Ask someone to find a _____ on the map. Now ask her to go to the _____. How many blocks does she have to walk?

| 1 | 2 | 3 | 4 | 5 | 6 | 7 | 8 | 9 | 10 | 11 | 12 |

Chapter 2 • 0 — 12

The Mail Carrier

Date _____

Directions: Mrs. Jackson is sorting the mail. How many letters altogether has she sorted?

IN TOWN   **OUT OF TOWN**

_____ + _____ = ☐

Answer:

◯ what you do when you read the word <u>altogether</u>.

+ or **−**

| 1 | 2 | 3 | 4 | 5 | 6 | 7 | 8 | 9 | 10 | 11 | 12 |

Chapter 2 • 0 — 12

Loading the Mail Truck

Date _____

Directions: Mr. O'Connor is loading big and little boxes on his truck for delivery. How many big and little boxes must he load altogether?

Hint: Word problems must have **labels** in the answer.

Answer:

_____ + _____ = ☐

Bonus: How many boxes would Mr. O'Connor have if he loaded only the little boxes?

_____

| 1 | 2 | 3 | 4 | 5 | 6 | 7 | 8 | 9 | 10 | 11 | 12 |

22 Chapter 2 • 0 — 12

Batting Practice

Date _____

Directions: Missy is practicing hitting balls before the game. How many baseballs did she hit altogether?

Answer:

_____ + _____ = ☐

Bonus: ◯ the word in the problem that tells you to add.

| 1 | 2 | 3 | 4 | 5 | 6 | 7 | 8 | 9 | 10 | 11 | 12 |

Chapter 2 • 0 — 12

Sports Photographer

Date _____

Directions: Mr. Gonzalez took photos of this week's high school basketball and volleyball games for the local newspaper. How many photos did he take in all?

Answer:

_____ + _____ = ☐

| 1 | 2 | 3 | 4 | 5 | 6 | 7 | 8 | 9 | 10 | 11 | 12 |

24 Chapter 2 • 0 — 12

Who Won?

Date _____

Directions: Add to find the final the scores. ◯ the winning team.

Team	1st	2nd	Final
Bulls	0	7	
Colts	3	6	

Show work:

Bulls	Colts

Hint: pts. is an abbreviation or short way for writing points.

| 1 | 2 | 3 | 4 | 5 | 6 | 7 | 8 | 9 | 10 | 11 | 12 |

Chapter 2 • 0 — 12

25

Bagging Baseball Equipment 1

Date _____

Directions: Directions: Coach Peterson made a list of equipment he needed for an away game. Use the list and the equipment already by the bag to solve the problems on the next page.

Coach's List

a. 10
b. 5
c. 7
d. 1
e. 12

| 1 | 2 | 3 | 4 | 5 | 6 | 7 | 8 | 9 | 10 | 11 | 12 |

26 Chapter 2 • 0 — 12

Bagging Baseball Equipment 2

Date _____

Directions: Solve to find out how many more of each item Coach Peterson needs to pack for the game. Show your work.

Hint: Remember to label your answers.

1 How many more 🏐 ?

2 How many more 🏏 ?

3 How many more 🥎 ?

4 How many more 🎭 ?

5 How many more ⛑ ?

6 Create your own problem.

| 1 | 2 | 3 | 4 | 5 | 6 | 7 | 8 | 9 | 10 | 11 | 12 |

Chapter 2 • 0 — 12

Throwing Darts

Date _____

Directions: Jesse throws darts in the recreation room. How many points altogether did he make?

____ + ____ + ____ = ☐ pts.

Bonus: Create another word problem using this page. Share your problem.

| 1 | 2 | 3 | 4 | 5 | 6 | 7 | 8 | 9 | 10 | 11 | 12 |

28

Chapter 2 • 0 — 12

Shooting Baskets

Date _____

Directions: Each player can shoot for the basket until she misses. Solve to find out how many baskets Tamara made.

Becky   **Sasha**   **Tamara**   **Total**

$$3 + 2 + \boxed{} = 8$$

Bonus: Create your own problem for some one else to solve.

| 1 | 2 | 3 | 4 | 5 | 6 | 7 | 8 | 9 | 10 | 11 | 12 |

Chapter 2 • 0 — 12

Tip Money

Date _____

Directions: Raven earned $6.00 in tips from her first three customers. Solve to find out how much money customer 1 tipped Raven.

Hint: Remember to use the **$** and **.** in your problem.

1. Customer 1

2. Customer 2

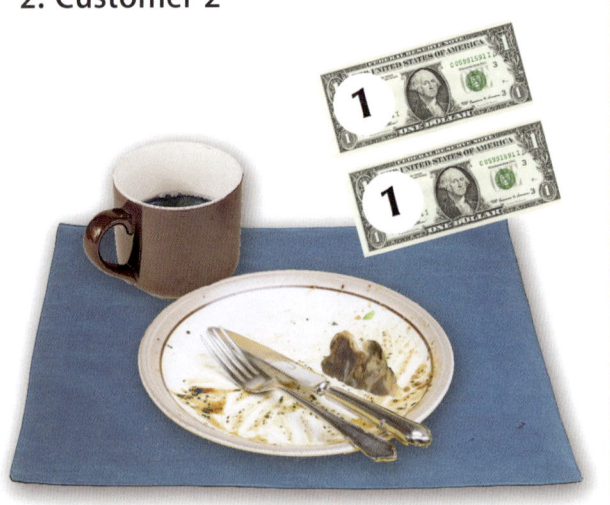

3. Customer 3

4. Total amount:

Show work:

1 2 3 4 5 6 7 8 9 10 11 12

Chapter 2 • 0 — 12

Day-Old Bake Sale

Date _____

Directions: Solve to find out how much money Roberta has left in her hand.

2.

3.

4.

Show work:

_____ ¢ – _____ ¢ = [___] ¢

Bonus: Create your own money problem using pennies.

| 1 | 2 | 3 | 4 | 5 | 6 | 7 | 8 | 9 | 10 | 11 | 12 |

Chapter 2 • 0 — 12

A Sale on Sunglasses

Date _____

Directions: Solve to find out how much money Dominick has left after he buys a pair of sunglasses.

Hint: Remember to use the **$** and **.** in your problem.

1.

2.

3.

4.

Show work:

_____ − _____ = ☐

1 2 3 4 5 6 7 8 9 10 11 12

Chapter 2 • 0 — 12

How Many Cookies Are in the Cookie Jar?

Date _____

Directions: Mrs. Thompson baked cookies for her children's school lunch. Solve to find out how many cookies are left to put in the cookie jar.

1. Mrs. Thompson baked 12 chocolate chip cookies.

2. She put 2 cookies in Manny's lunch.

3. She put 2 cookies in each twin's lunch.

4. How many cookies were left to put in the cookie jar?

Show work:

_____ − _____ = ☐

| 1 | 2 | 3 | 4 | 5 | 6 | 7 | 8 | 9 | 10 | 11 | 12 |

Chapter 2 • 0 — 12

A Trip to the Mall 1

Date _____

Directions: Use the prices on this page to solve the problems on the next page.

- CDs $6.00
- shorts $8.00
- dresses $7.00
- tees $5.00
- back packs $10.00
- caps $3.00

Bonus: When you read the word <u>left</u> do you + or − ?

Chapter 2 • 0 — 12

A Trip to the Mall 2

Date _____

Directions: ✘ out the money you spent. ◯ how much money you have left.

Problem: Show work:

1. You have: (9 one-dollar bills)   You bought: (UB-50 Live! CD)   What do you have left? (9 one-dollar bills)

2. You have: (12 one-dollar bills)   You bought: (backpack)   What do you have left? (12 one-dollar bills)

3. You have: (12 one-dollar bills)   You bought: (shirt and cap)   What do you have left? (12 one-dollar bills)

4. Create your own problem. You have: You bought: What do you have left?

Challenge: Use a calculator to find out how much each purchase cost with a 5% sales tax.

| 1 | 2 | 3 | 4 | 5 | 6 | 7 | 8 | 9 | 10 | 11 | 12 |

35

Chapter 2 • 0 — 12

Saturday Practice

Date _____

Directions: Draw hands on the analog clock to match the digital time.

1 Mary goes to swim practice Saturday mornings at 10:00.

2 Stacy goes to practice her golf swing Saturday afternoons at 12:00.

1 2 3 4 5 6 7 8 9 10 11 12

36 Chapter 2 • 0 — 12

Guitar Practice

Date _____

Directions: Draw hands on the analog clocks to match the digital time. Solve the problem.

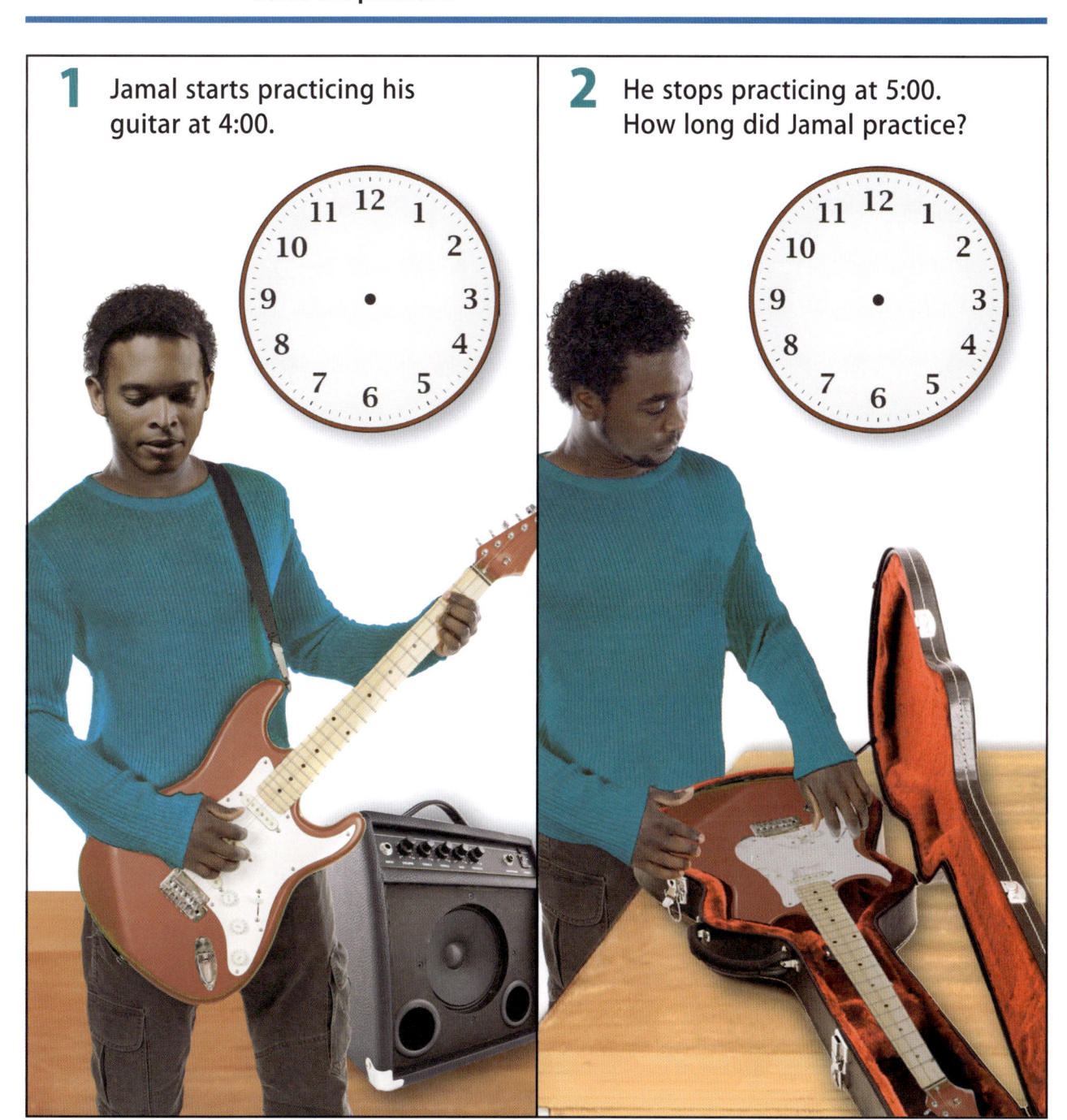

1 Jamal starts practicing his guitar at 4:00.

2 He stops practicing at 5:00. How long did Jamal practice?

Bonus: Jamal practices his guitar every day after school. How many hours a week does Jamal practice?

| 1 | 2 | 3 | 4 | 5 | 6 | 7 | 8 | 9 | 10 | 11 | 12 |

Chapter 2 • 0 — 12

The Mail Carrier

Date _____

Directions: Listen to your teacher read the problem.
Draw hands on the analog clocks to solve the problem

1 Mr. Jackson starts his route at 8:00 a.m.

2 He stops to eat a packed lunch 3 hours later. At what time does Mr. Jackson eat lunch?

Bonus: If you eat lunch at 12:00 and school started 4 hours earlier, at what time did you start school?

| 1 | 2 | 3 | 4 | 5 | 6 | 7 | 8 | 9 | 10 | 11 | 12 |

Chapter 2 • 0 — 12

A Daily Calendar

Date _____

Directions: Chance is busy today. He needs to refer to his calendar to be sure he remembers everything he has to do. Use the calendar to answer the questions below.

Tuesday— April 6, 2010

Time	Activity		Time	Activity
7:00	Breakfast at school		12:00	Lunch
			12:30	Study Hall
8:00	Math class		1:00	Computer class
9:00	Art class		2:00	Choir
10:00	English class		3:00	Soccer Practice
11:00	Science class		4:00	Dentist appt.
			5:00	Walk the dog

Answer:

1. When does Chance have computer class? _____
2. When does Chance have English class? _____
3. How many hours from the time Chance eats breakfast until lunch? _____
4. If Chance is at soccer practice, how many hours ago was he in science class? _____
5. How many hours from the time Chance went to the dentist until he walked his dog? _____

Bonus: How long was Chance's study hall?

| 1 | 2 | 3 | 4 | 5 | 6 | 7 | 8 | 9 | 10 | 11 | 12 |

Chapter 2 • 0 — 12

Mrs. Ling's Calendar 1

Date _____

Directions: Mrs. Ling is the high school principal.
Use her busy schedule to answer the questions on the next page.

October 2010

Sunday	Monday	Tuesday	Wednesday	Thursday	Friday	Saturday
10	11	12	13	14	15	16
	Columbus Day	school board mtg.	girls' volleyball game	orchestra concert	homecoming	Sweetest Day

1 2 3 4 5 6 7 8 9 10 11 12

40

Chapter 2 • 0 — 12

Mrs. Ling's Calendar 2

Date _____

Directions: Use the calendar to answer the questions.

1. How many days in a week? _____

2. ◯ the 1st day of the week on the calendar.

3. Put a ✔ on the last day of the week.

4. Put a ☆ star on the 3rd day of the week.

5. What happens on the 5th day of the week? (circle one)

6. What happens on the 4th day of the week? (circle one)

7. What happens on the 6th day of the week? (circle one)

8. Point to Sunday. How many days before Mrs. Ling must talk to the school board? _____

9. Point to Tuesday. How many days before homecoming? _____

10. What do you think Mrs. Ling's husband will get her for Sweetest Day? (circle one)

Bonus: Create your own problem using Mrs. Ling's calendar.

| 1 | 2 | 3 | 4 | 5 | 6 | 7 | 8 | 9 | 10 | 11 | 12 |

Chapter 2 • 0 — 12

A Yearly Calendar 1

Date _____

Directions: Mr. Jefferson is a TV producer. Help him remember his busy schedule for the year. Listen to your teacher read his calendar.

January shoot episode for detective show	**February** Super Bowl	**March** H.S. basketball tournament
April shoot episodes for Reality TV	**May** Music Awards	**June** Wild animal special
July 4th of July	**August** Reality TV begins	**September** Football begins
October Detective show begins	**November** American Superstar	**December** Space Odyssey

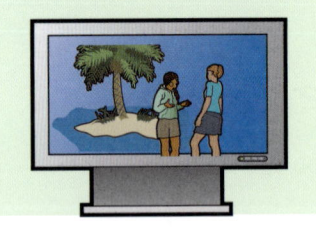

Chapter 2 • 0 — 12

A Yearly Calendar 2

Date _____

Directions: Use Mr. Jefferson's calendar to answer the questions.

1. Point to and say the months in order.

2. In what month is the 🏈 ? **February** or **March**

3. In what month is the 🦁 ? **June** or **July**

4. What show does Mr. Jefferson produce in March? 🏈 or 🏀

5. What show does Mr. Jefferson produce in November? ⭐ or 🏆

6. It is June. How many months before the 🪐 is shown? _____

7. It is January. How many months before the 📺 is shown? _____

8. Mr. Jefferson is producing the 🏆.
   How many months before he produces the ⭐? _____

9. The 📺 starts in August. How many months ago did Mr. Jefferson begin shooting the show? _____

Bonus: Create a problem using Mr. Jefferson's calendar.

| 1 | 2 | 3 | 4 | 5 | 6 | 7 | 8 | 9 | 10 | 11 | 12 |

Chapter 2 • 0 — 12

Penny Candy

Date _____

Directions: ◯ the coins that match the total price.

Purchase	Coins you have
1.  3¢	
2.  4¢	
3.  11¢	
4.  7¢	
5. 3¢ + 3¢	
6. 7¢ + 3¢	

1 2 3 4 5 6 7 8 9 10 11 12

44 Chapter 2 • 0 — 12

Penny Candy Again

Date _____

Directions: ◯ the correct change in each box.

You have	You buy	Your change
1. (nickel)	candy 3¢	5 pennies
2. (nickel)	candy stick 4¢	5 pennies
3. (2 nickels)	mint chocolate 8¢	5 pennies
4. (2 nickels)	kiss 7¢	5 pennies
5. (2 nickels)	3¢ + 3¢	5 pennies
6. (2 nickels)	chocolate sweet milk bar 9¢	5 pennies

1 2 3 4 5 6 7 8 9 10 11 12

Chapter 2 • 0 — 12

45

Mr. Jefferson's TV Shows 1

Date _____

Directions: The TV network asked people which one of Mr. Jefferson's shows they liked best. Use the graph to answer the questions on the next page about the top-rated shows.

Popular TV Shows

VIEWERS	Super Bowl	Space Odyssey	Music Awards	Reality TV	Wild Animal Special
12					
11					
10	●				
9	●				●
8	●				●
7	●			●	●
6	●			●	●
5	●			●	●
4	●	●		●	●
3	●	●		●	●
2	●	●	●	●	●
1	●	●	●	●	●
0	●	●	●	●	●

46 Chapter 2 • 0 — 12

Mr. Jefferson's TV Shows 2

Date _____

Directions: Use the graph on p. 46 to solve the problems below.

1. Which show has the most viewers?

   football or lion

   (circle one)

2. Which show has the least amount of viewers?

   space or music

   (circle one)

3. How many people altogether watched

   island and music ?

   _____

4. How many people altogether watched

   space and music ?

   _____

5. How many more people watched

   lion than island ?

   _____

6. How many more people watched

   football than space ?

   _____

1   2   3   4   5   6   7   8   9   10   11   12

47

Chapter 2 • 0 — 12

Warm-Up Exercises 1

Date _____

Directions: Coach Peterson's team uses some exercises to warm up before the game. Use the Venn diagram below to answer the questions on the next page.

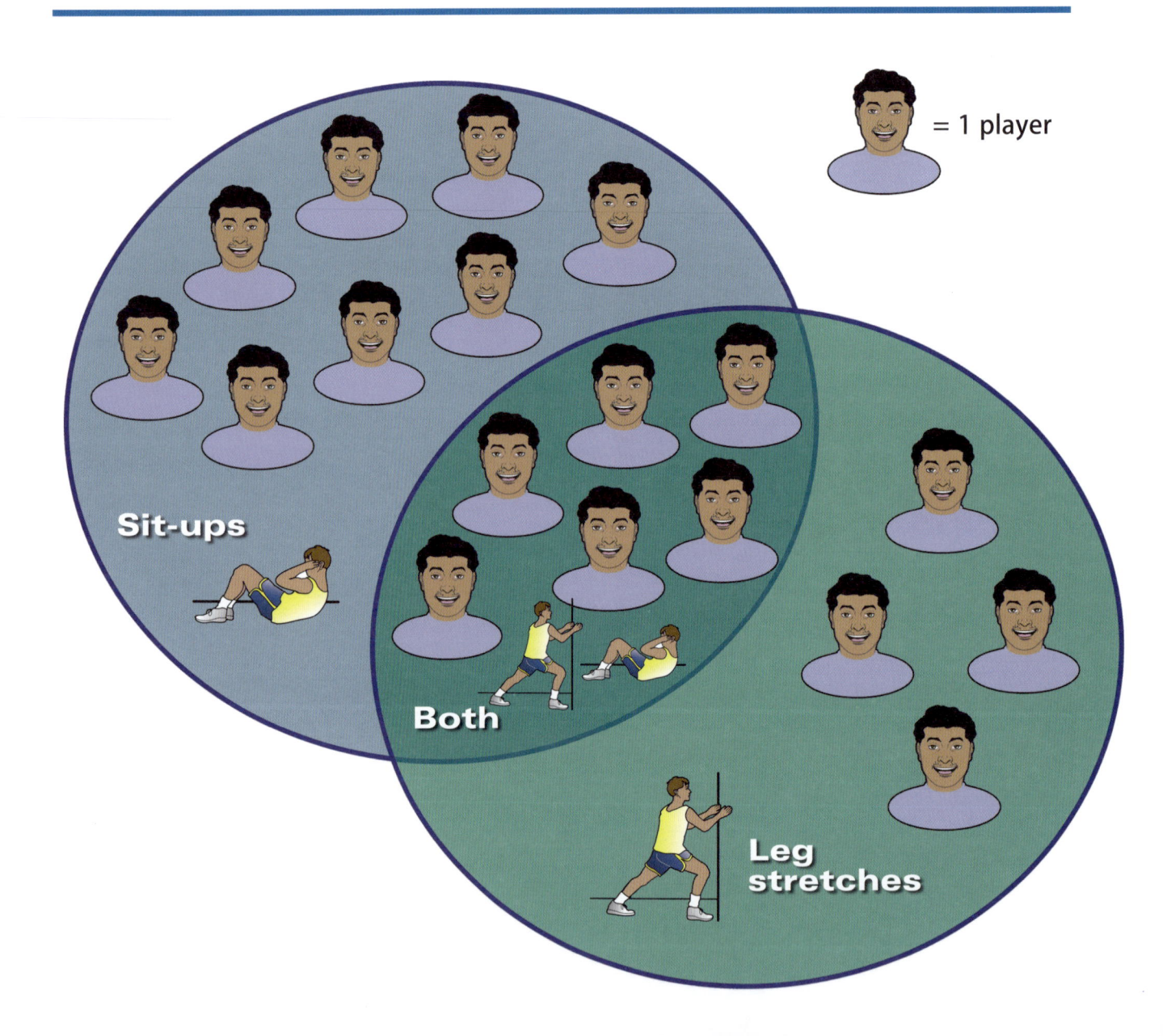

= 1 player

Sit-ups

Both

Leg stretches

Key

1. Sit-ups:
2. Leg stretches:
3. Both: ___ and ___

48 Chapter 2 • 0 — 12

Warm-Up Exercises 2

Date _____

Directions: Use the graph on p. 48 to solve the problems below.

1 How many players do sit-ups?

2 How many players do leg stretches?

3 How many players do both?

4 How many more players do sit-ups than do  leg stretches?

5 How many more players do both exercises than do leg stretches?

6 Create your own problem.

1　2　3　4　5　6　7　8　9　10　11　12

Chapter 2 • 0 — 12

New Tiger Cubs

Date _____

Directions: The Bengal tiger in the zoo had cubs.
Use all of the animals to answer the questions below.

Answer:

1. If 1 tiger has 1 tail, how many tails will 4 tigers have? _____

2. If 1 tiger has 2 ears, how many ears will 3 tigers have? _____

3. If I tiger has 4 legs, how many legs will 3 tigers have? _____

4. Challenge:
   If 1 tiger has 2 eyes, how many eyes will 5 tigers have? _____

| 1 | 2 | 3 | 4 | 5 | 6 | 7 | 8 | 9 | 10 | 11 | 12 |

Chapter 2 • 0 — 12

Coin Flip

Date _____

Directions: Get a penny. Predict whether the coin will land more on heads or tails. ◯ heads or tails below.
Flip the coin 12 times. Mark a tally on the chart each time you flip the coin.
Count the tallies. Was your prediction right?

Prediction:

heads   or   tails

Tally Chart

Heads	Tails
Total: _____	Total: _____

Prediction: _____ **yes**   or   _____ **no**

Chapter 2 • 0 — 12

Chapter 3
0 — 18

How Many Moves?

Date _____

Directions: Solve to find out how many spaces the player can move her game piece on the board below.

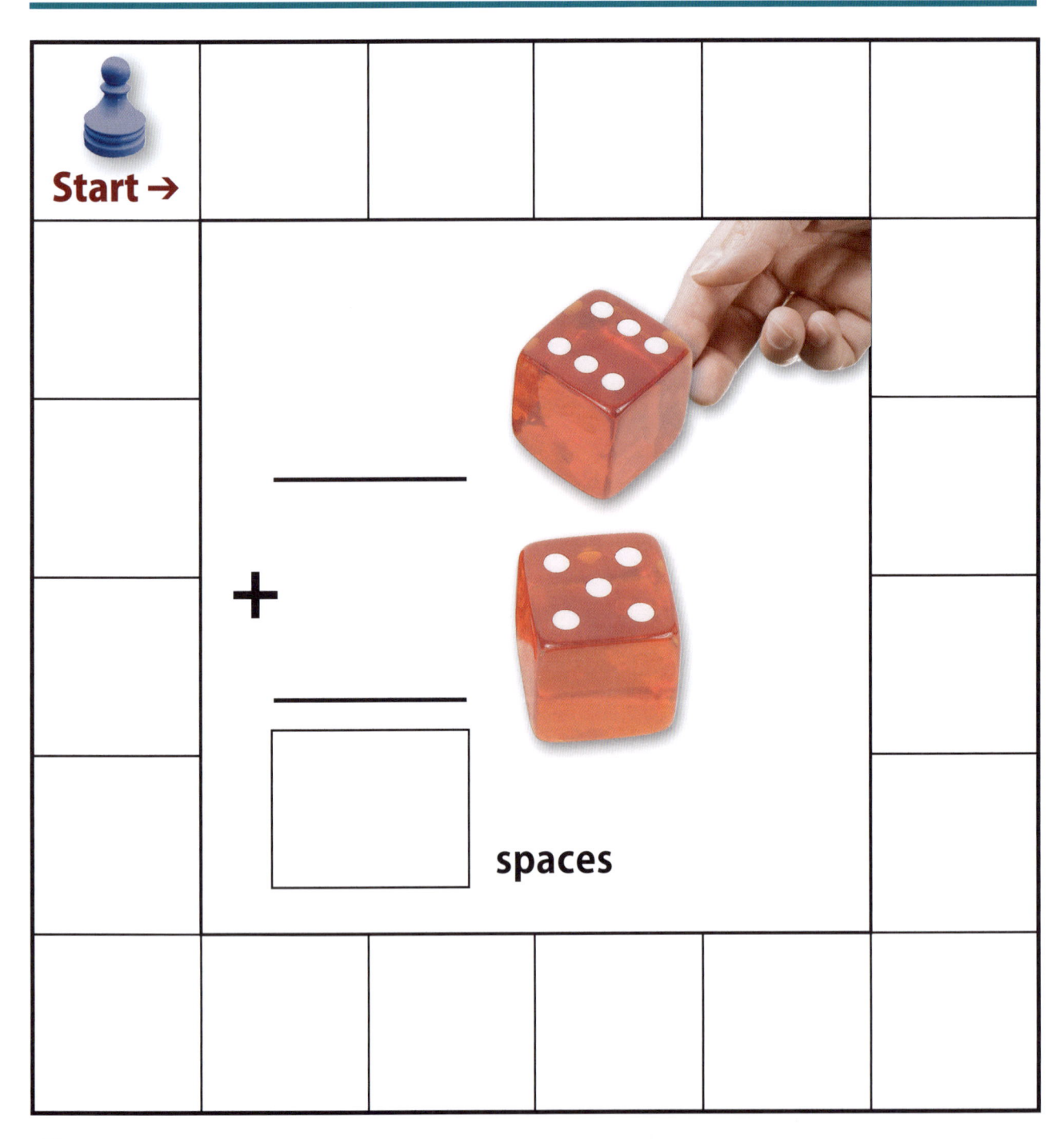

_____
+
_____

spaces

Bonus: Shade the number of spaces on the game board to match the numbers on the dice.

1 2 3 4 5 6 7 8 9 10 11 12 13 14 15 16 17 18

54

Chapter 3 • 0 — 18

A Walk About the Neighborhood

Date _____

Directions: Shanice is going to walk to several places in her town. Trace the path that Shanice walks as you answer the questions below.

Answer:

1. ◯ the 🏢 county courthouse.

2. Shanice goes to 🏠 Abhay's house. How many blocks does Shanice walk? _____ blocks

3. Shanice and Abhay walk to the 🏛 library. How many blocks do they walk? _____ blocks

4. Shanice says good-bye to Abhay and starts to walk home. She stops at the clothing store to get a sweater. How many blocks did she walk from the library to the clothing store? _____ blocks.

5. Put a check on 🏠 Shanice's house. How many more blocks did Shanice walk from the clothing store? _____

Bonus: How many blocks in all did Shanice walk? _____ blocks

| 1 | 2 | 3 | 4 | 5 | 6 | 7 | 8 | 9 | 10 | 11 | 12 | 13 | 14 | 15 | 16 | 17 | 18 |

Pick Your Own

Date _____

Directions: Maria went with her family to pick apples at an apple orchard. How many apples did Maria pick altogether?

$+$ ___

Bonus: Maria gave 3 🍎 to her brother. How many does she have left? _____

1 2 3 4 5 6 7 8 9 10 11 12 13 14 15 16 17 18

Chapter 3 • 0 — 18

Wheelbarrows Full of Pumpkins

Date _____

Directions: Mrs. Thompson just finished picking pumpkins to sell in her roadside stand. How many big and little pumpkins did Mrs. Thompson pick altogether?

Solve:

_____ + _____ = _____ s

| 1 | 2 | 3 | 4 | 5 | 6 | 7 | 8 | 9 | 10 | 11 | 12 | 13 | 14 | 15 | 16 | 17 | 18 |

Chapter 3 • 0 — 18

Roadside Stand 1

Date _____

Directions: Mrs. Thompson runs a roadside stand at the apple orchard. Use this page to solve the problems on the next 3 pages.

Hint: lbs. is a short way or abbreviation for writing pounds.

- Apples: 5 lb. — $5.00
- Onions: 2 lb. — $2.00
- Potatoes: 10 lb. — $4.00
- Small Pumpkins: 3 lb. — $3.00
- Carving Pumpkins: 6 lb. — $6.00

58

Chapter 3 • 0 — 18

Roadside Stand 2

Date _____

Directions: Use the picture on page 58 to find out how many pounds (lbs.) of fruit and vegetables each customer bought. Show all work.

Problem:	Show work:
1. Customer 1:   5 lb. + 10 lb. = _____	
2. Customer 2:   6 lb. + 3 lb. = _____	
3. Customer 3:   10 lb. + 2 lb. = _____	
4. Customer 4:   3 lb. + 5 lb. = _____	

1 2 3 4 5 6 7 8 9 10 11 12 13 14 15 16 17 18

Chapter 3 • 0 — 18

Roadside Stand 3

Date _____

Directions: Use the prices found on page 58. ◯ the amount of money each customer must pay.

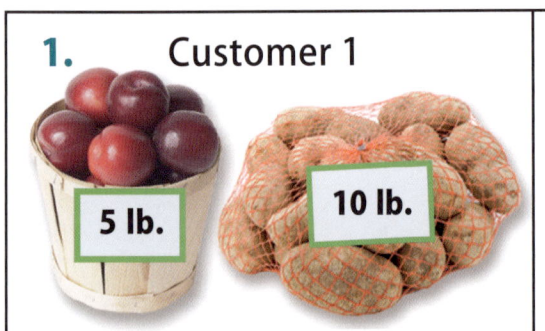

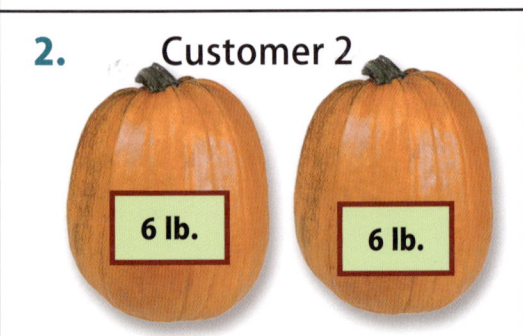

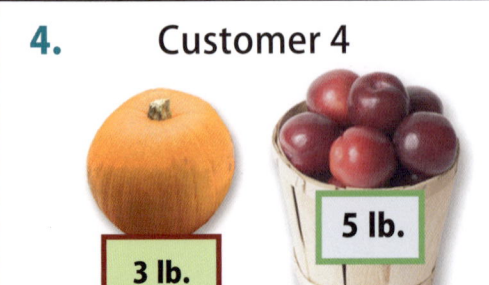

Problem:	Total amount:
1. Customer 1 — 5 lb., 10 lb.	1 1 1 1 / 1 1 1 1 / 1 1 1 1
2. Customer 2 — 6 lb., 6 lb.	1 1 1 1 / 1 1 1 1 / 1 1 1 1
3. Customer 3 — 10 lb., 2 lb.	1 1 1 1 / 1 1 1 1
4. Customer 4 — 3 lb., 5 lb.	1 1 1 / 1 1 1 / 1 1 1

Bonus: Create your own problem using the produce on page 58.

1 2 3 4 5 6 7 8 9 10 11 12 13 14 15 16 17 18

60

Chapter 3 • 0 — 18

Roadside Stand 4

Date _____

Directions: Use the prices found on page 58. Write the cost of each item and solve for the total amount the customer must pay.

Hint: Remember to use the **$** and **.** The first one is done for you.

1 Roadside Stand Customer 1

Item	Amount
1. 5 lb.	
2. 2 lb.	
Total	

2 Roadside Stand Customer 2

Item	Amount
1. 6 lb.	
2. 3 lb.	
Total	

3 Roadside Stand Customer 3

Item	Amount
1. 10 lb.	
2. 2 lb.	
Total	

4 Roadside Stand Customer 4

Item	Amount
1. 3 lb.	
2. 5 lb.	
Total	

Bonus: ◯ the customer who spent the most money. Put a ✓ on the customer who spent the least amount of money.

1 2 3 4 5 6 7 8 9 10 11 12 13 14 15 16 17 18

Chapter 3 • 0 — 18

A Hockey Win

Date _____

Directions: Manny missed the second half of the game. Solve to find the missing scores.

Team	1st	2nd	Final
HOME	5		11
AWAY	3		5

Show work:

HOME:

AWAY:

1 2 3 4 5 6 7 8 9 10 11 12 13 14 15 16 17 18

Chapter 3 • 0 — 18

Photo Album

Date _____

Directions: Solve to find out how many photos Tammy had in her album before she took 5 additional photos.

1. Tammy loves to take photos and put them in her scrapbook. She had some photos in her book.

2. Tammy took 5 more photos. She put them in her book.

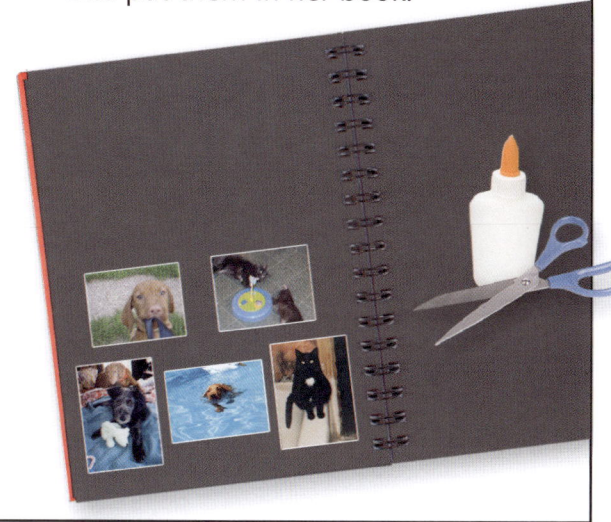

3. Tammy now has 18 photos in her book.

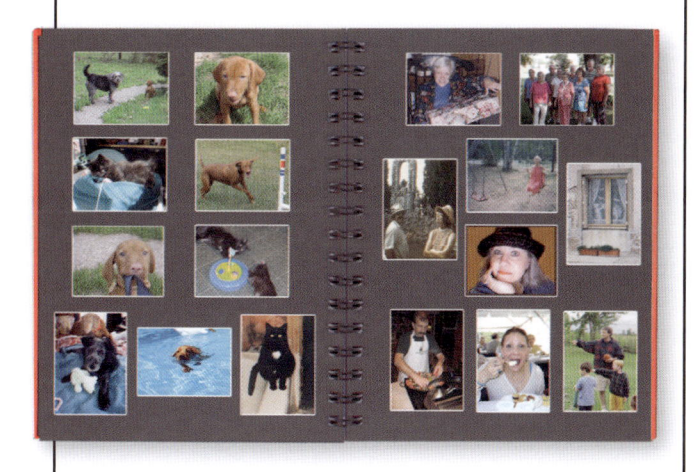

4. How many photos did Tammy originally have in her album?

Show work:

1 2 3 4 5 6 7 8 9 10 11 12 13 14 15 16 17 18

Chapter 3 • 0 — 18

63

A Scrabble Game

Date _____

Directions: Solve to find the total points for each word. Don't forget to label your answer. **Hint:** A short way or abbreviation for writing points is **pts.**

Player	Show work:
1 Player 1  — $F_4 O_1 X_8$	
2 Player 2 — $B_3 A_1 D_2$ 	
3 Player 3 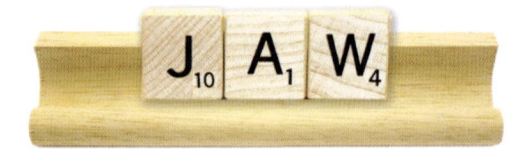 — $J_{10} A_1 W_4$	
4 Player 4 — $M_3 A_1 N_1 Y_4$ 	
5 Player 5  — $Q_{10} U_1 I_1 E_1 T_1$	

Bonus: Which player has the most points? Which player has the least?

1 2 3 4 5 6 7 8 9 10 11 12 13 14 15 16 17 18

Chapter 3 • 0 — 18

A Weekend Trip

Date _____

Directions: How many miles did the Kilpatrick family travel each day?
Hint: Remember to label your answer. A short way to write miles is **mi.**

Day One

The Kilpatrick family started at Tyrellpass and stopped for the night in Castledaly.

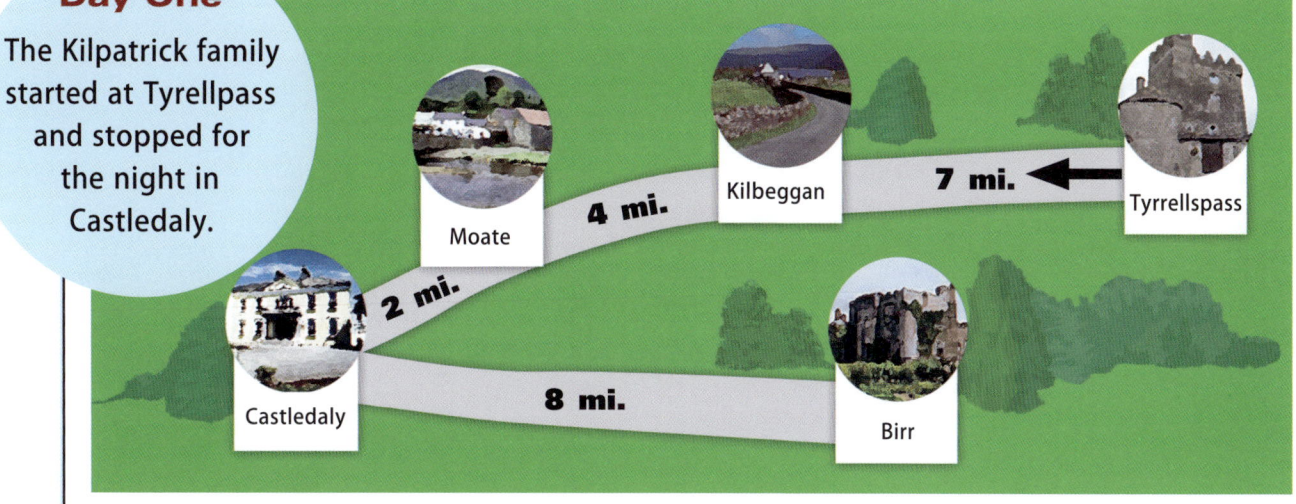

_____ mi. **+** _____ mi. **+** _____ mi. **=** ☐ _____

Day Two

The Kilpatrick family left Casteldaly and returned to Tyrellpass using a different route.

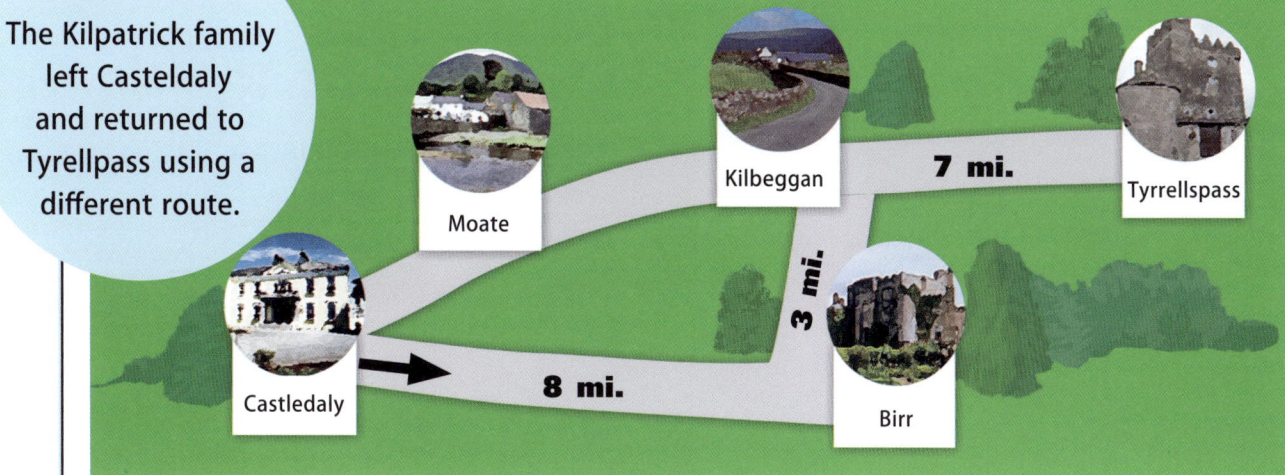

_____ mi. **+** _____ mi. **+** _____ mi. **=** ☐ _____

1 2 3 4 5 6 7 8 9 10 11 12 13 14 15 16 17 18

Chapter 3 • 0 — 18

Tamika's Free Throws

Date _____

Directions: Coach Peterson is recoding the amount of points Tamika made at the free throw line in the first 3 games. Solve to find out how many free throw points Tamika made in the 3rd game.

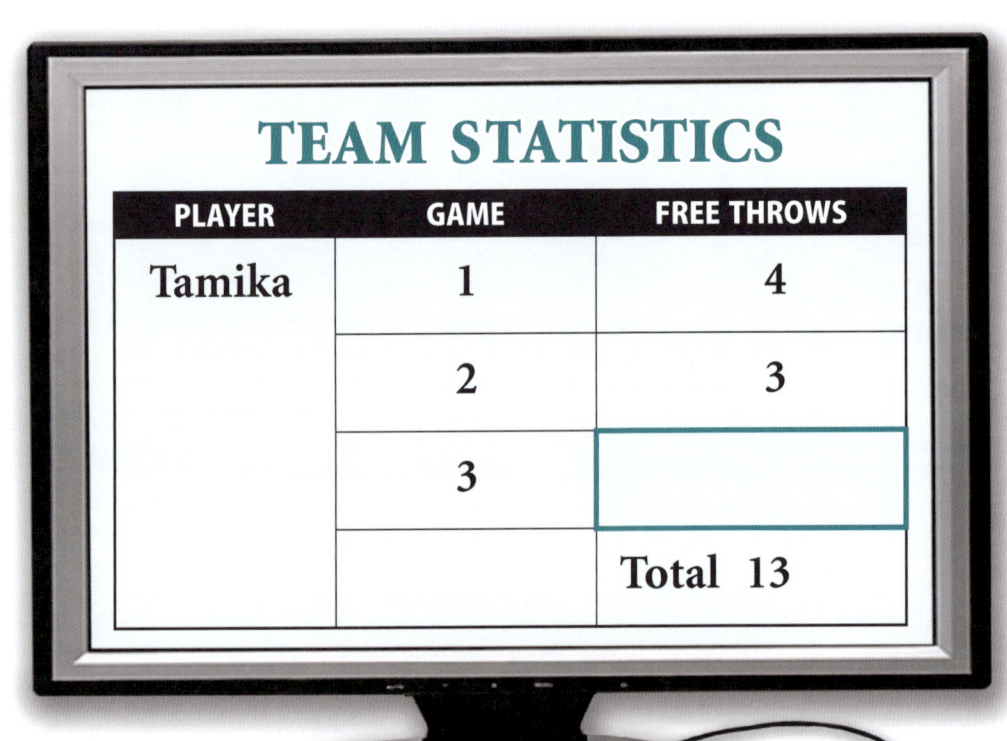

TEAM STATISTICS

PLAYER	GAME	FREE THROWS
Tamika	1	4
	2	3
	3	
		Total 13

Show work:

1 2 3 4 5 6 7 8 9 10 11 12 13 14 15 16 17 18

Chapter 3 • 0 — 18

Stocking Shelves

Date _____

Directions: How many boxes of cereal are left on the cart after Henry stocked the shelves?
Hint: Remember to label your answer.

1 Henry saw that there were no cereal boxes left on the shelves.

2 He loaded the cart to stock the shelves with cereal boxes.

3 Henry put 8 boxes on the shelves.

4 How many cereal boxes did Henry have left on the cart?

Show work:

1 2 3 4 5 6 7 8 9 10 11 12 13 14 15 16 17 18

Chapter 3 • 0 — 18

Home vs. Away

Date _____

Directions: Solve to find out how many more points the Home team won by.
Hint: Remember to label your answer.

Team	1st	2nd	FINAL
HOME	10	7	17
AWAY	7	6	13

Show work:

1 2 3 4 5 6 7 8 9 10 11 12 13 14 15 16 17 18

68

Chapter 3 • 0 — 18

BR-R-R-R It's Cold!

Date _____

Directions: Solve to find the difference between the high and low temperatures.
Hint: ° is a short way of writing degrees.

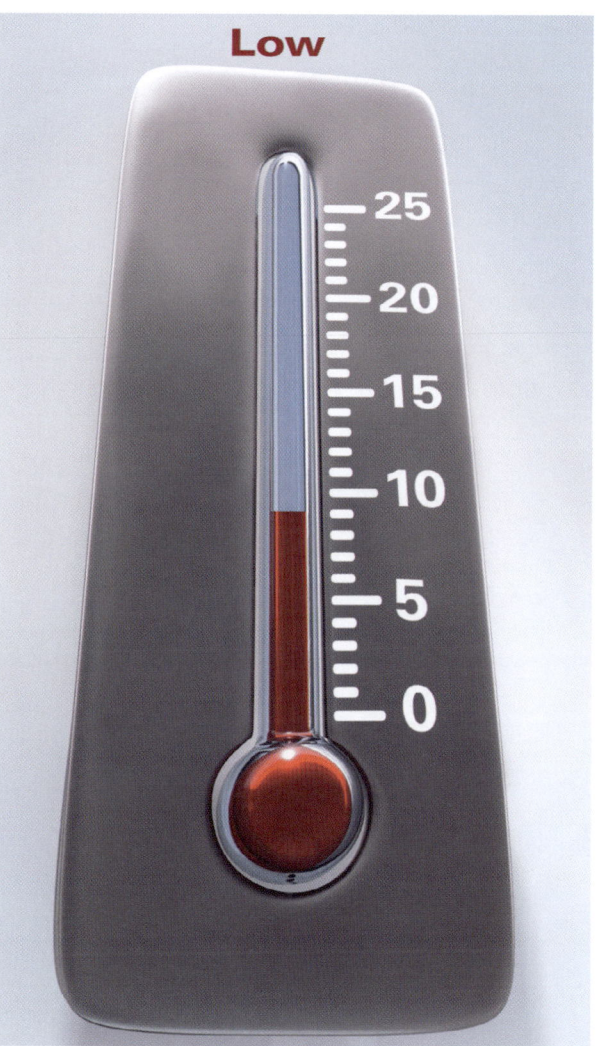

High

Low

Show work:

_____ − _____ = ☐

Bonus: If the red goes up on the thermometer, it is: **warmer** or **colder** (circle one)

1 2 3 4 5 6 7 8 9 10 11 12 13 14 15 16 17 18

Chapter 3 • 0 — 18

A Seven-Day Winter Forecast 1

Date _____

Directions: Use this page to solve the problems on the next page.

Sunday	Monday	Tuesday	Wednesday	Thursday	Friday	Saturday
High: 18	High: 16	High: 9	High: 12	High: 17	High: 17	High: 15
Low: 10	Low: 8	Low: 2	Low: 7	Low: 8	Low: 11	Low: 12

Wear warm clothes this week. It will be cold!

Trace to 0.5"

Trace to 0.5"

0.5" to 2"

Trace Amounts

COLD!

1 2 3 4 5 6 7 8 9 10 11 12 13 14 15 16 17 18

70 Chapter 3 • 0 — 18

A Seven-Day Winter Forecast 2

Date _____

Directions: Find the difference between the high and low temperatures for each day.
Hint: ° is a short way of writing degrees. Don't forget the label!

Show work:

1. Sunday	2. Monday	3. Tuesday
4. Wednesday	5. Thursday	6. Friday
7. Saturday	8. ◯ the day with the lowest temperature. Put an ✗ on the day with the highest temperature.	

1 2 3 4 5 6 7 8 9 10 11 12 13 14 15 16 17 18

Chapter 3 • 0 — 18

The State Fair 1

Date _____

Directions: Solve to find out how many bottles were knocked over.

1. Softball Toss	2. Softball Toss	3. Show work:

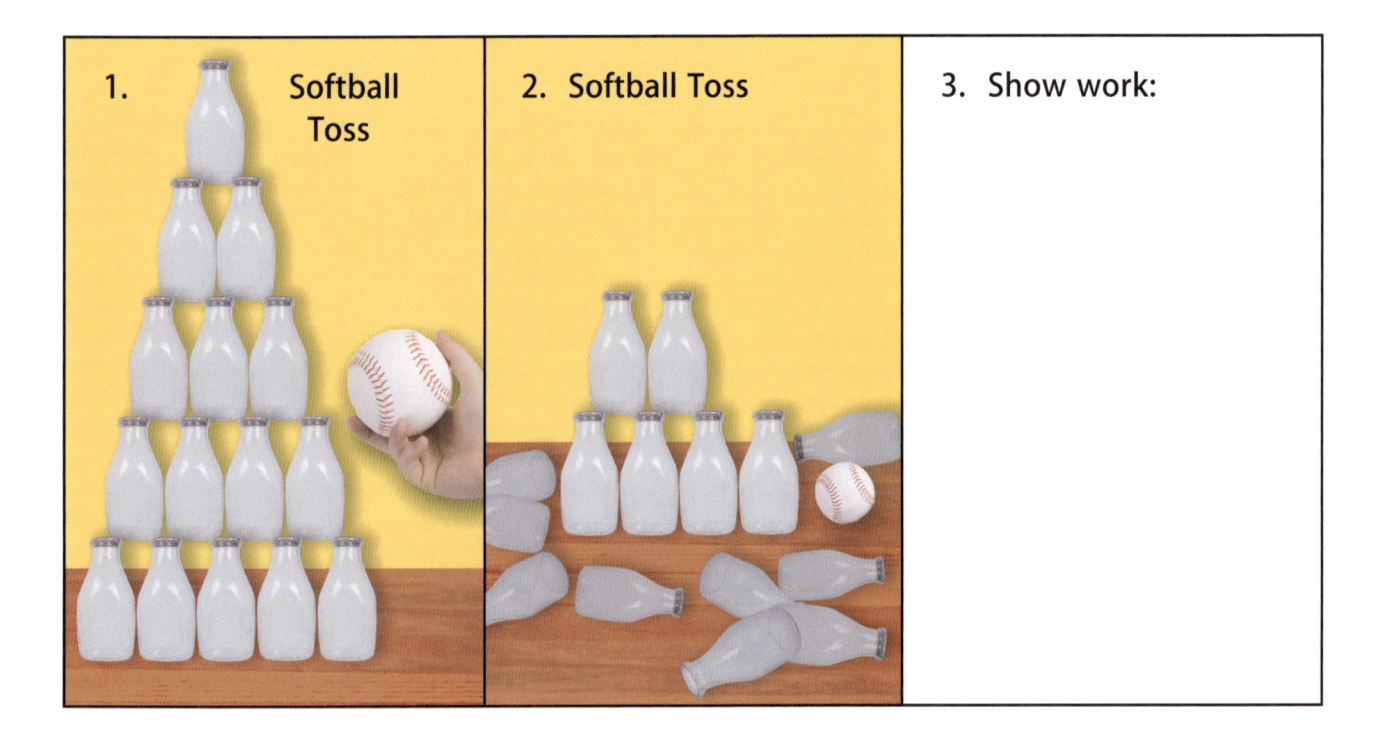

Directions: Solve to find out how many balloons were popped.

1. Dart Throw	2. Dart Throw	3. Show work:

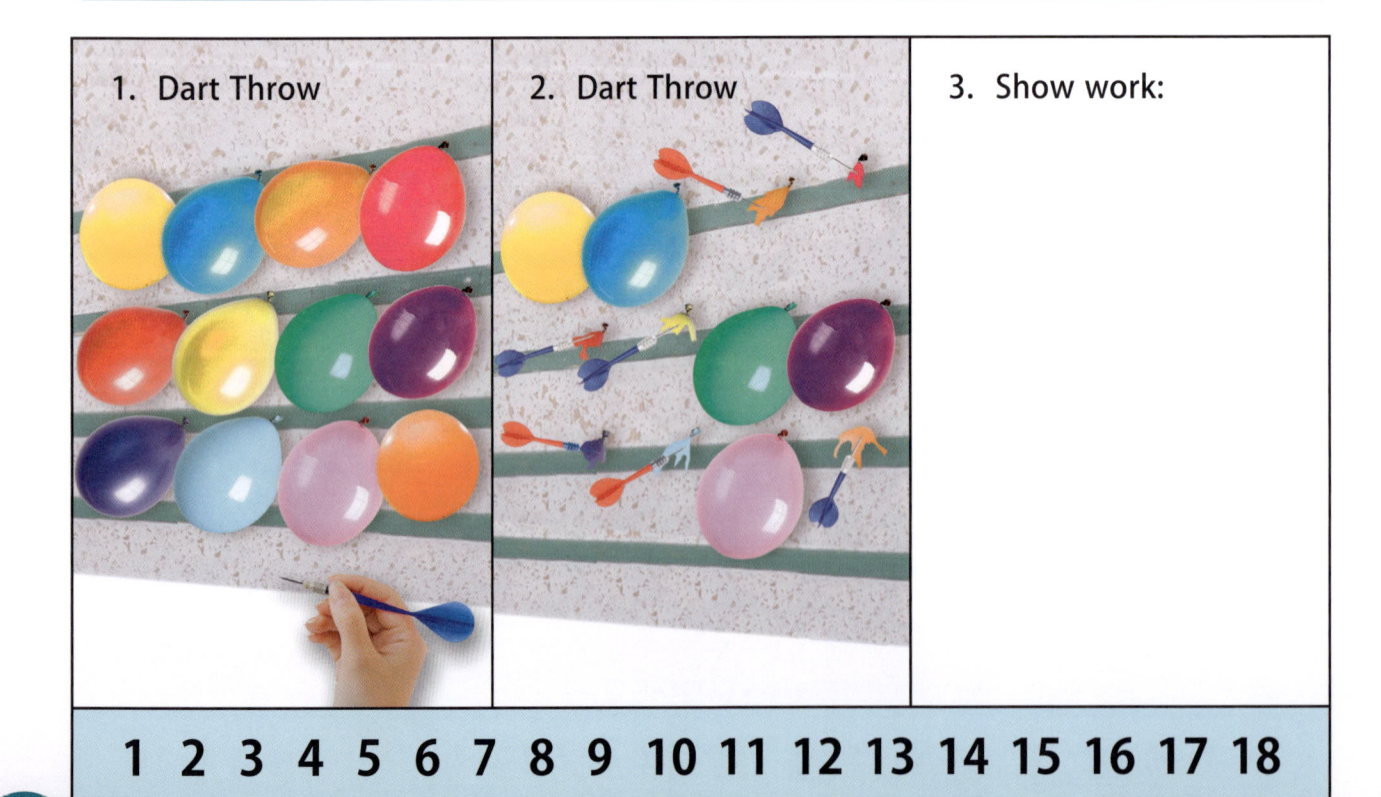

1 2 3 4 5 6 7 8 9 10 11 12 13 14 15 16 17 18

Chapter 3 • 0 — 18

The State Fair 2

Date _____

Directions: Solve to find our how many tickets Melissa has left.

1. Melissa bought 14 tickets.

2. She gave 3 tickets to her brother.

3. She gave 2 more tickets to her friend.

4. How many tickets does Melissa have left?

Show work:

1 2 3 4 5 6 7 8 9 10 11 12 13 14 15 16 17 18

Chapter 3 • 0 — 18

73

I Missed the First Half!

Date _____

Directions: Solve to find how many points each team made during the 1st half. Write the score for each team in that box.
Hint: A short way of writing points is **pts.**

Team	1st	2nd	FINAL
HOME		7	14
AWAY		14	17

Show work:

Home

Away

Bonus: ◯ the team that won the game.   HOME   AWAY

1 2 3 4 5 6 7 8 9 10 11 12 13 14 15 16 17 18

74

Chapter 3 • 0 — 18

Team Statistics

Date _____

Directions: Listen to your teacher read the problem.
Draw hands on the analog clocks and solve the problem

1. Coach Lebowski finished the statistics for each player at 8:00.

2. He started 4 hours earlier. What time did the coach start working?

_____

Bonus: Create your own time problem.

1 2 3 4 5 6 7 8 9 10 11 12 13 14 15 16 17 18

Chapter 3 • 0 — 18

Walking the Dog

Date _____

Directions: Listen to your teacher read the problem.
Draw hands on the analog clocks and solve the problem.

1. Melinda started to walk her dog at 3:00.

2. She stopped walking him a half an hour later. When did Melinda stop walking her dog?

   _____

1 2 3 4 5 6 7 8 9 10 11 12 13 14 15 16 17 18

76 Chapter 3 • 0 — 18

A Librarian at Work

Date _____

Directions: Listen to your teacher read the problem.
Draw hands on the analog clocks and solve the problem..

1. Mrs. Yang started to check in books that were returned to the library at 1:30.

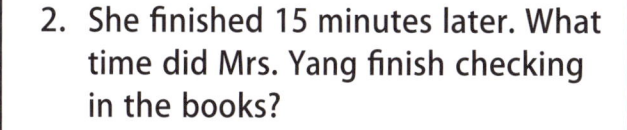

2. She finished 15 minutes later. What time did Mrs. Yang finish checking in the books?

_____

Answer: If it is 12:45, what time will it be in 15 more minutes? _____

| 1 | 2 | 3 | 4 | 5 | 6 | 7 | 8 | 9 | 10 | 11 | 12 | 13 | 14 | 15 | 16 | 17 | 18 |

Chapter 3 • 0 — 18

77

A Weekend Job 1

Date _____

Directions: Listen to your teacher read the problem.
Draw hands on the analog clocks and solve the problem.

Saturday:

1. Tomas clears the sidewalks for some of his neighbors after it snows. He started at 9:00.

2. He finished 3 hours later. What time did he stop working?

   _____

Sunday:

3. Tomas worked again on Sunday. He started at 10:00.

4. He finished 2 and 1/2 hours later. What time did he stop working?

   _____

Bonus: How many hours did Tomas work?

| 1 | 2 | 3 | 4 | 5 | 6 | 7 | 8 | 9 | 10 | 11 | 12 | 13 | 14 | 15 | 16 | 17 | 18 |

78

Chapter 3 • 0 — 18

A Weekend Job 2

Date _____

Directions: Listen to your teacher read the problem. ◯ the amount of money Tomas earned each day clearing the sidewalks of snow.

Saturday:

1. Tomas gets $3.00 for each job.

2. Today he had 6 jobs. How much did he earn?

Sunday:

3. Tomas gets $4.00 a job on Sundays.

4. Today he had 4 jobs. How much did he earn?

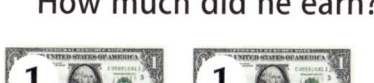

Bonus: How much did he earn for the weekend?

1 2 3 4 5 6 7 8 9 10 11 12 13 14 15 16 17 18

Chapter 3 • 0 — 18

Flea Market 1

Date _____

Directions: ◯ the coins that match the price of the item bought.

Purchase: | **Coins you have:**

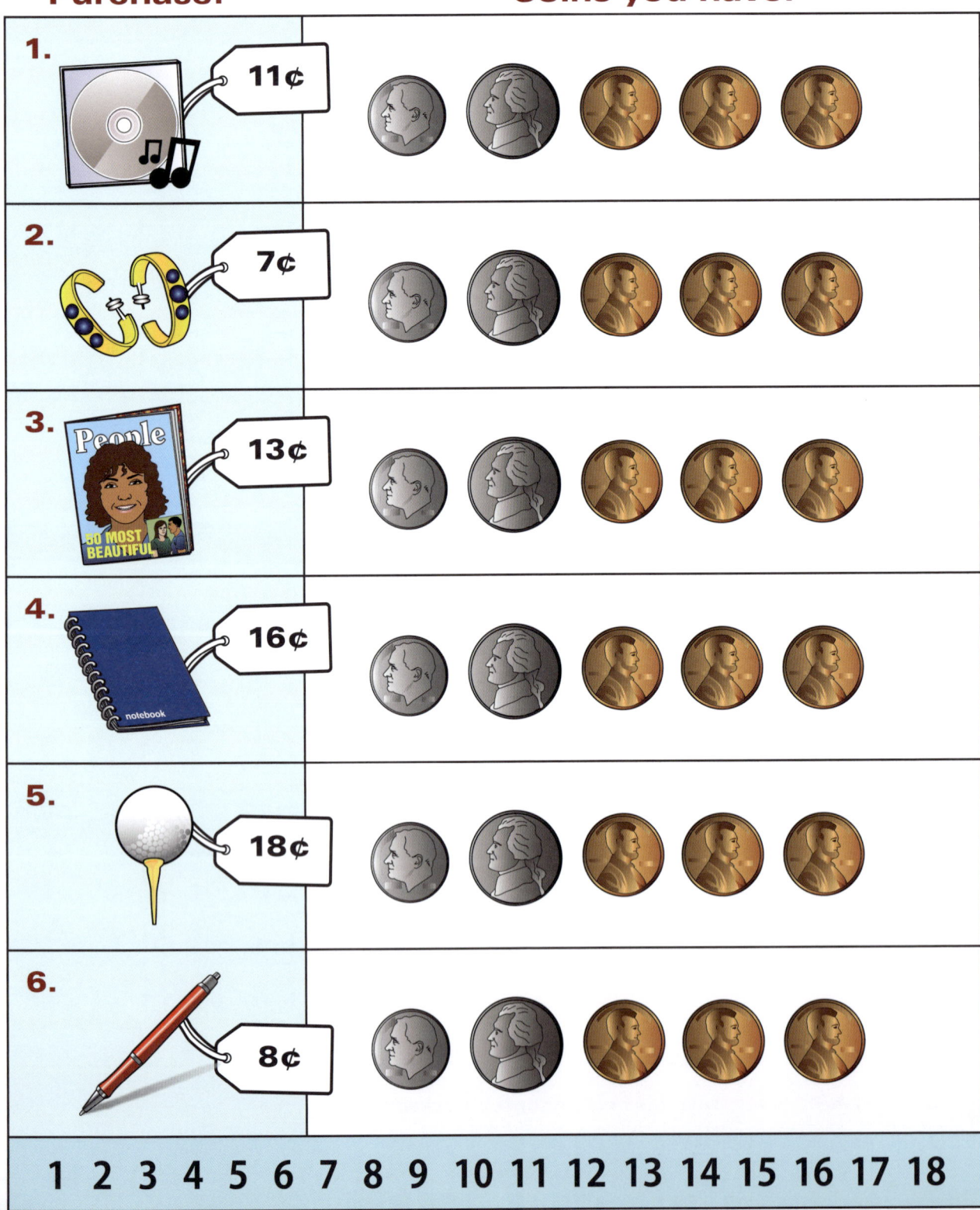

1. 11¢
2. 7¢
3. 13¢
4. 16¢
5. 18¢
6. 8¢

1 2 3 4 5 6 7 8 9 10 11 12 13 14 15 16 17 18

80   Chapter 3 • 0 — 18

Flea Market 2

Date _____

Directions: ◯ the correct change in each box.

You have:	You buy:	Your change:
1.	11¢ (CD)	
2.	8¢ (notebook)	
3.	13¢ (People magazine)	
4.	4¢ (earrings)	
5.	2¢ + 2¢ (paint)	
6.	8¢ + 4¢ (sketchbook and pencil)	

1 2 3 4 5 6 7 8 9 10 11 12 13 14 15 16 17 18

81

Chapter 3 • 0 — 18

Sports News 1

Date _____

Directions: Graph the wins for each team on the next page.

Weekly Gazette
Home Town Teams Win Big This Season

Undefeated!

Boys Win!

Cross-country Goes to State!

Statistics	
Team	Wins
Cross-country	12
Boy's Volleyball	15
Girl's Volleyball	17
Football	10

Girls Tie for First

1 2 3 4 5 6 7 8 9 10 11 12 13 14 15 16 17 18

Chapter 3 • 0 — 18

Sports News 2

Date _____

Directions: Use the statistics chart on the previous page to graph the wins for each team.

No. of wins

Sports News

18
16
14
12
10
8
6
4
2
0

Teams: cross-country | boy's volleyball | girl's volleyball | football

Bonus: How many more games have the girl's volleyball team won than the cross-country team?

1 2 3 4 5 6 7 8 9 10 11 12 13 14 15 16 17 18

Taking Inventory 1

Date _____

Directions: Part of Henry's job as a stock boy is to count the number of items on the shelf. Add the total amount for each item and use it to complete the graph on the next page.

Hint: A short way of writing number is **no.**

Inventory

Item	No. on the shelf	Total
1. chicken soup	||||| |||||		
2. canned spaghetti	||||| ||||| ||||| |||	
3. tuna fish	||||| ||	
4. macaroni	||||| ||||| |||||		
5. pizza mix	||||| ||||| ||||	

Bonus: How many more cans of chicken soup does Henry have to stock to have 16 cans on the shelf?

| 1 | 2 | 3 | 4 | 5 | 6 | 7 | 8 | 9 | 10 | 11 | 12 | 13 | 14 | 15 | 16 | 17 | 18 |

84

Chapter 3 • 0 — 18

Taking Inventory 2

Date _____

Directions: Use the Tally Chart on the previous page to graph the total amount of items on the shelves in the grocery store.

Taking Inventory

No. of items on the shelf: 0, 2, 4, 6, 8, 10, 12, 14, 16, 18

Items: Soup, Spaghetti, Tuna, Macaroni & Cheese, Pizza

Bonus: ◯ the column with the least amount.

How many more pizza mixes are there than packages of tuna?

1 2 3 4 5 6 7 8 9 10 11 12 13 14 15 16 17 18

Chapter 3 • 0 — 18

Cafeteria Orders

Date _____

Directions: Solve to find out how many orders were made.

1. 7 students each ordered 2 hot dogs. How many hot dogs were ordered?

   Answer: _____

2. 6 students each ordered 3 cheese pizza slices. How many cheese pizzas were ordered?

   Answer: _____

3. 4 students each ordered 2 cheeseburgers for themselves and 2 cheeseburgers for a friend. How many cheeseburgers were ordered?

   Answer: _____

Use a calculator to find out how much food was ordered in all.

Answer: _____

1 2 3 4 5 6 7 8 9 10 11 12 13 14 15 16 17 18

Chapter 3 • 0 — 18

Track Stars

Date _____

Directions: Find the track star with blue numbers on his uniform.
Write **Y** if it is true and **N** if it is false.

Color	Alex	Tony	Kareem
red			
blue			
black			

1. Alex's numbers are red.
2. Kareem's numbers are not blue.
3. ◯ the track star with blue numbers.

Directions: Find the track star that does the broad jump.
Write **Y** if it is true and **N** if it is false.

Event	Tamika	Alexandra	Isabel
hurdles			
pole vault			
broad jump			

1. Alexandra does the hurdles.
2. Tamika does not do the broad jump.
3. ◯ the track star that does the broad jump.

1 2 3 4 5 6 7 8 9 10 11 12 13 14 15 16 17 18

Chapter 3 • 0 — 18

Team	1st	2nd	TOTAL
STATE		40	83
AWAY		22	72

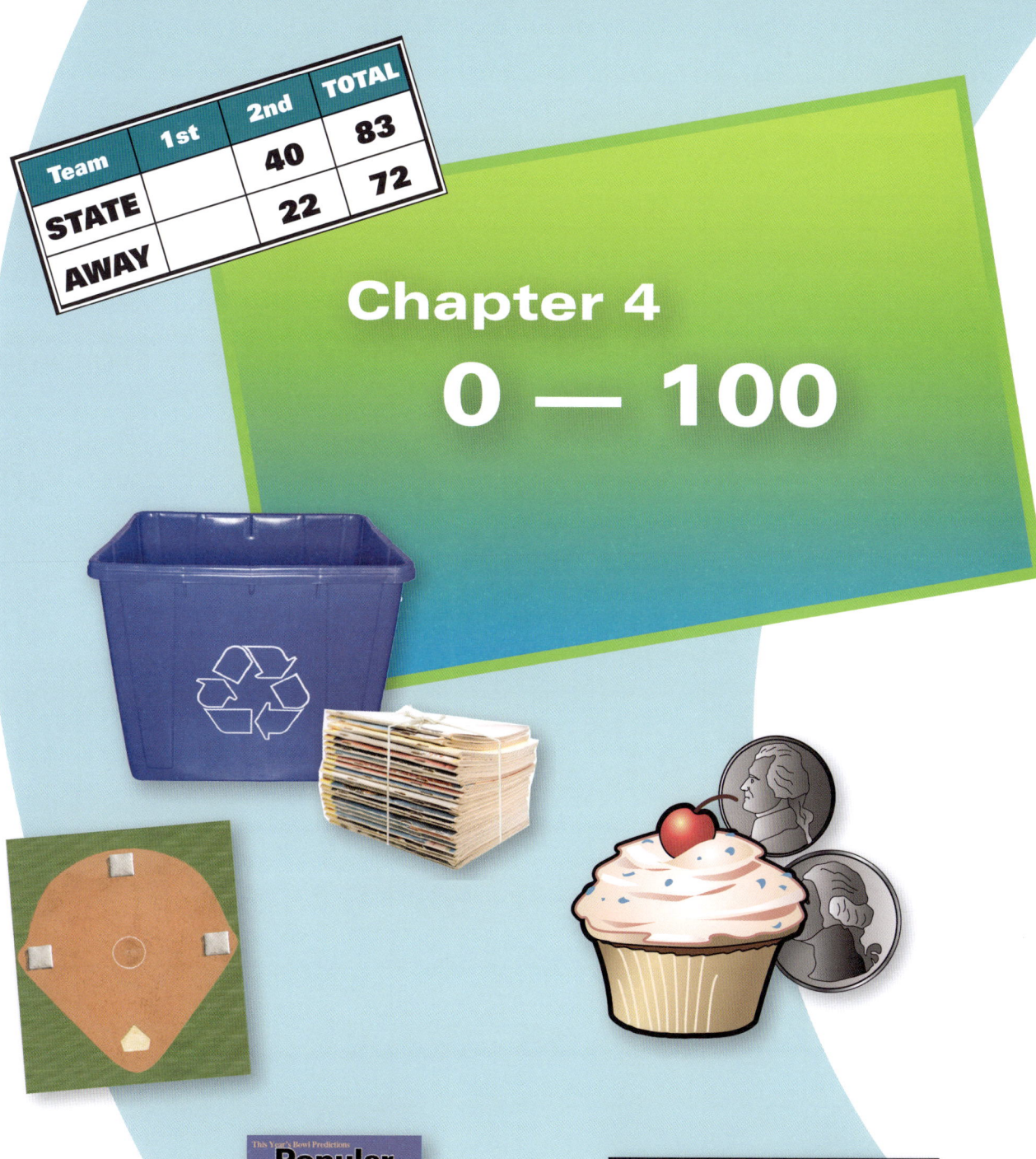

Chapter 4
0 — 100

51	67	78
8:00 a.m. (morning)	12:00 (noon)	5:00 p.m. (afternoon)

Chapter 4 • 0 — 100

One Hundreds Chart

1	2	3	4	5	6	7	8	9	10
11	12	13	14	15	16	17	18	19	20
21	22	23	24	25	26	27	28	29	30
31	32	33	34	35	36	37	38	39	40
41	42	43	44	45	46	47	48	49	50
51	52	53	54	55	56	57	58	59	60
61	62	63	64	65	66	67	68	69	70
71	72	73	74	75	76	77	78	79	80
81	82	83	84	85	86	87	88	89	90
91	92	93	94	95	96	97	98	99	100

Counting Change

Date _____

Directions: ◯ the amount of money needed to pay for the snack.

Hint: Try to use the least amount of change.

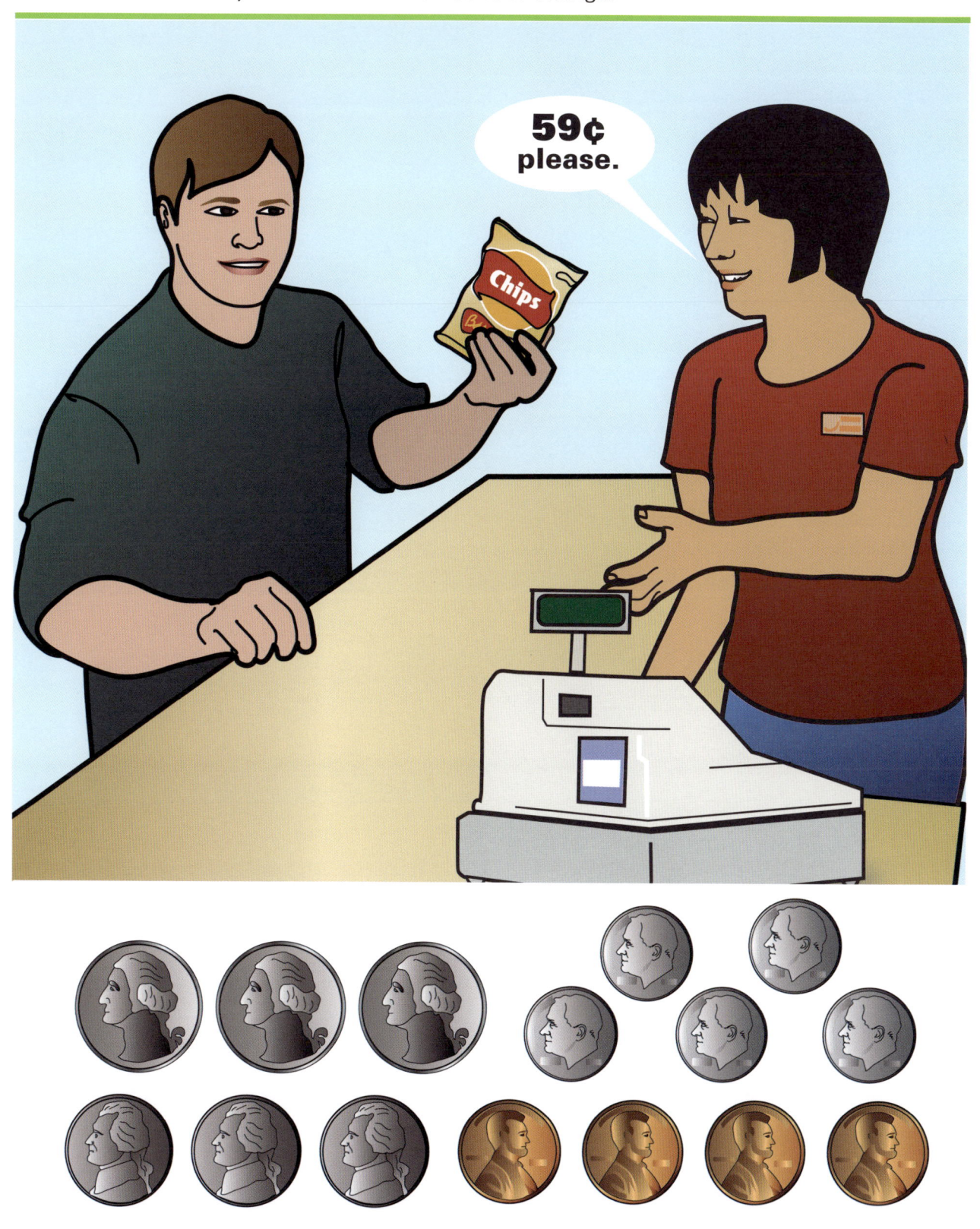

59¢ please.

Chapter 4 • 0 — 100

91

Center City

Date _____

Directions: Answer the questions below.

Key
- School
- Mall
- City Hall
- Post Office
- Museum
- Park

1. ⭕ city hall.

2. Put an **X** on the museum.

3. Put a ✔ on the high school.

4. How many blocks from city hall to the museum? _____ blocks

5. How many blocks from the high school to the middle school? _____ blocks

6. How many blocks from the park to the high school. _____ blocks

7. Ask someone a question about the map.

Bonus: Make a map of one floor of your school.

Chapter 4 • 0 — 100

New Softball Field

Date _____

Directions: Coach Peterson is going to redesign the softball field.

Using a ruler, draw home plate, 1st base, 2nd base, and 3rd base where they belong on the field.

Make sure that home plate and the 3 bases are 1 inch long and 1 inch wide squares. Label the bases.

◯ the pitcher's mound.

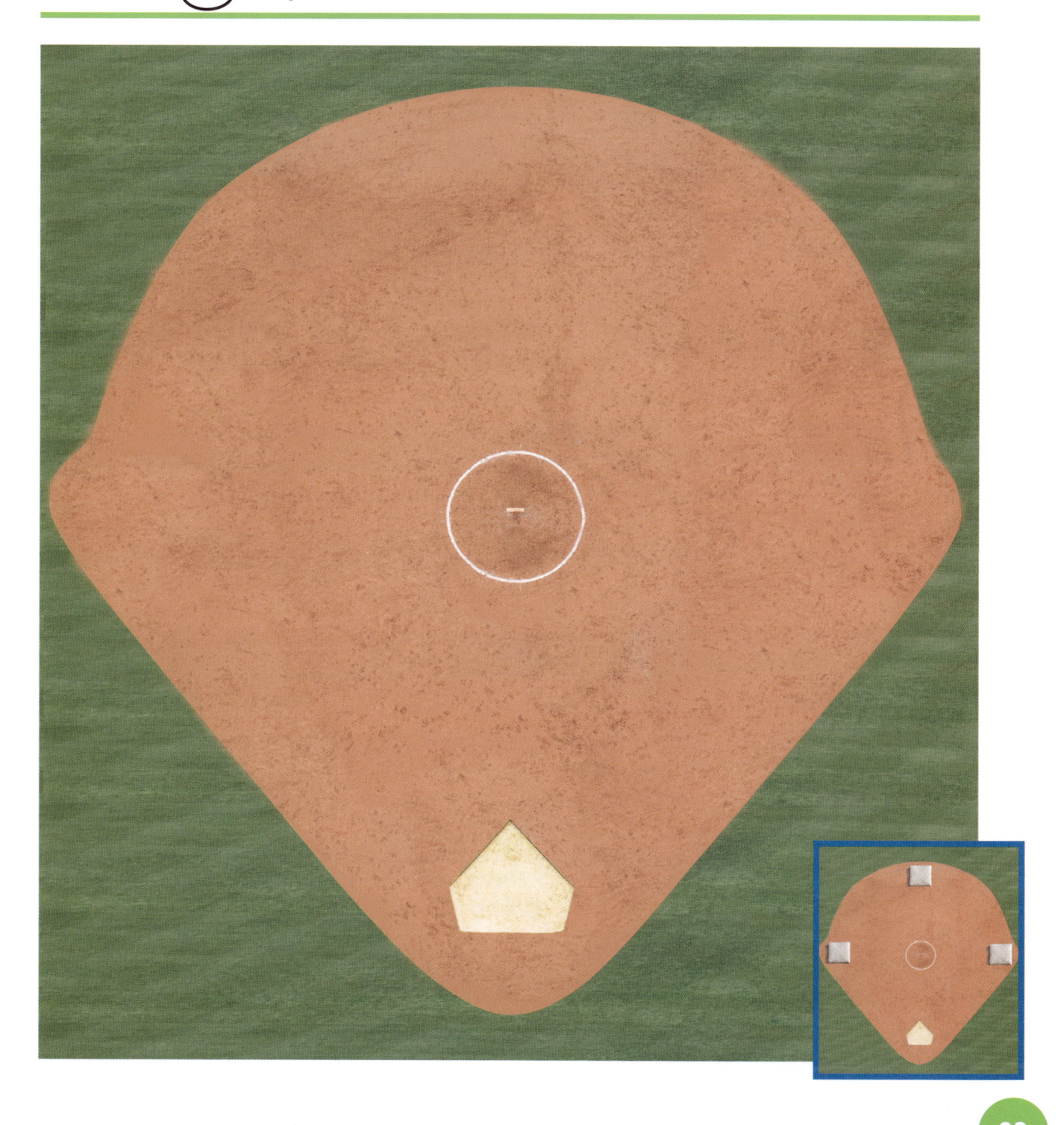

Band Bake Sale 1

Date _____

Directions: Use the Bake Sale menu to solve the problems on the next page.

Central High Band Boosters

Item	Amount
cookie	15¢
muffin	25¢
cupcake	30¢
pie	35¢
cake	40¢

Sale!

Band Bake Sale 2

Date _____

1 ○ the number of 🍪 you can buy if you have 🪙🪙.

2 ○ the coins you would use if you bought a 🧁.

3 ○ what you could buy if you have 🪙🪙.

4 ○ what you could buy if you have 🪙.

5 ○ the coins you would use if you bought  🥧 and 🧁.

6. Challenge: ○ the change you would get back if you bought a 🍪 and had this much money 🪙 🪙 🪙.

Chapter 4 • 0 — 100

High-Rise Office Building 1

Date _____

Directions: Use the elevator buttons to find out how many floors each person rode in the elevator.

Chapter 4 • 0 — 100

High-Rise Office Building 2

Date _____

Directions: Use the elevator buttons on p. 72 to solve the problems.
Hint: **L** is the first floor or floor **1** of the building.

1 How many floors are there in this office building?

2 Kimberly got on at floor **9** and off at floor **36**. How many floors did she ride?

3 Mr. Sykes got on at **L**. He went to his office on floor **24**. How many floors did he ride?

4 Tamara is on floor **18**. She wants to go to **32**. How many floors must she ride?

5 Mrs. Ling is on floor **25**. She pushes the button for floor **29**. How many floors will she ride?

6 Challenge: Alejandro got on at floor **15** and off on floor **2**. How many floors did he ride?

Recycling Newspapers and Magazines

Date _____

Directions One of Liam's jobs in the high-rise office building is to bundle newspapers and magazines to recycle. Solve to find out how many newspapers and magazines he put in the recycle bin.

Hint: = 10 newspapers   = 1 newspaper

   = 10 magazines   = 1 magazine

+

98   Chapter 4 • 0 — 100

Sandwiches for Sale

Date _____

Directions: Mrs. Lao prepares sandwiches to sell in the high-rise office building cafeteria. Solve to find out how many sandwiches Mrs. Lao made to sell.

Hint: 🍔 or 🥪 = 10

_____ 🍔 + _____ 🥪 = ☐

Bonus: Each sandwich cost $2.00. How many sandwiches can you buy if you have this much money? _____

Chapter 4 • 0 — 100

A Business Meeting

Date _____

Directions: Ms. Kennedy is helping Mr. Sykes prepare for a business meeting in the high-rise office building. Every person who sits down must have a handout. How many more handouts does Ms. Kennedy have to place on the chairs?

Hint: Remember to label your answer.

Solve:

Find the Missing Scores

Date _____

Directions: Find and write the missing scores for each game.

Show work:

Game 1

Team	1st	2nd	TOTAL
STATE		25	68
AWAY		13	39

Game 2

Team	1st	2nd	TOTAL
STATE		40	83
AWAY		22	72

Game 3

Team	1st	2nd	TOTAL
STATE		34	85
AWAY		42	97

Game 4

Team	1st	2nd	TOTAL
STATE		36	67
AWAY		44	66

Bonus:
How many games did State win? _____ lose? _____

Chapter 4 • 0 — 100

Summer Sports 1

Date _____

Directions: Coach Feinstein runs a summer sports program. Use the numbers on the clipboard to answer the questions on the next page.

Summer Sports

Sport	Students
1. Soccer	88
2. Golf	62
3. Softball	95
4. Basketball	77
5. Swimming	44
6. Tennis	23

Summer Sports 2

Date _____

Directions: Use page 102 to solve the problems below.

1 ◯ the most popular sport.

2 How many more students play  🏀 than 🏌️?

3 How many more students play ⚽ than 🥽?

4 How many more students play 🏀 than 🎾?

5 How many students in all play 🎾 and 🏌️?

6 Coach Feinstein wants 26 students on the 🎾 team. How many more students have to sign up?

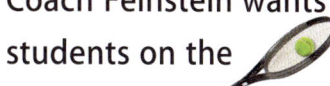

Chapter 4 • 0 — 100

103

Recording Highs and Lows

Date _____

Directions: Tamara Jefferson is the meteorologist for Channel 15 and must take the highs and lows of the day for the weather forecast. Read the thermometers and write the temperature on the clipboard. Solve to find the difference between the high and low temperatures.

High   **Low**

Solve:

104 Chapter 4 • 0 — 100

A Day's Forecast

Date _____

Directions: Tamara gives today's forecast.
Use the forecast to answer the questions below.
Hint: your answer must have ° (degrees) as a label.

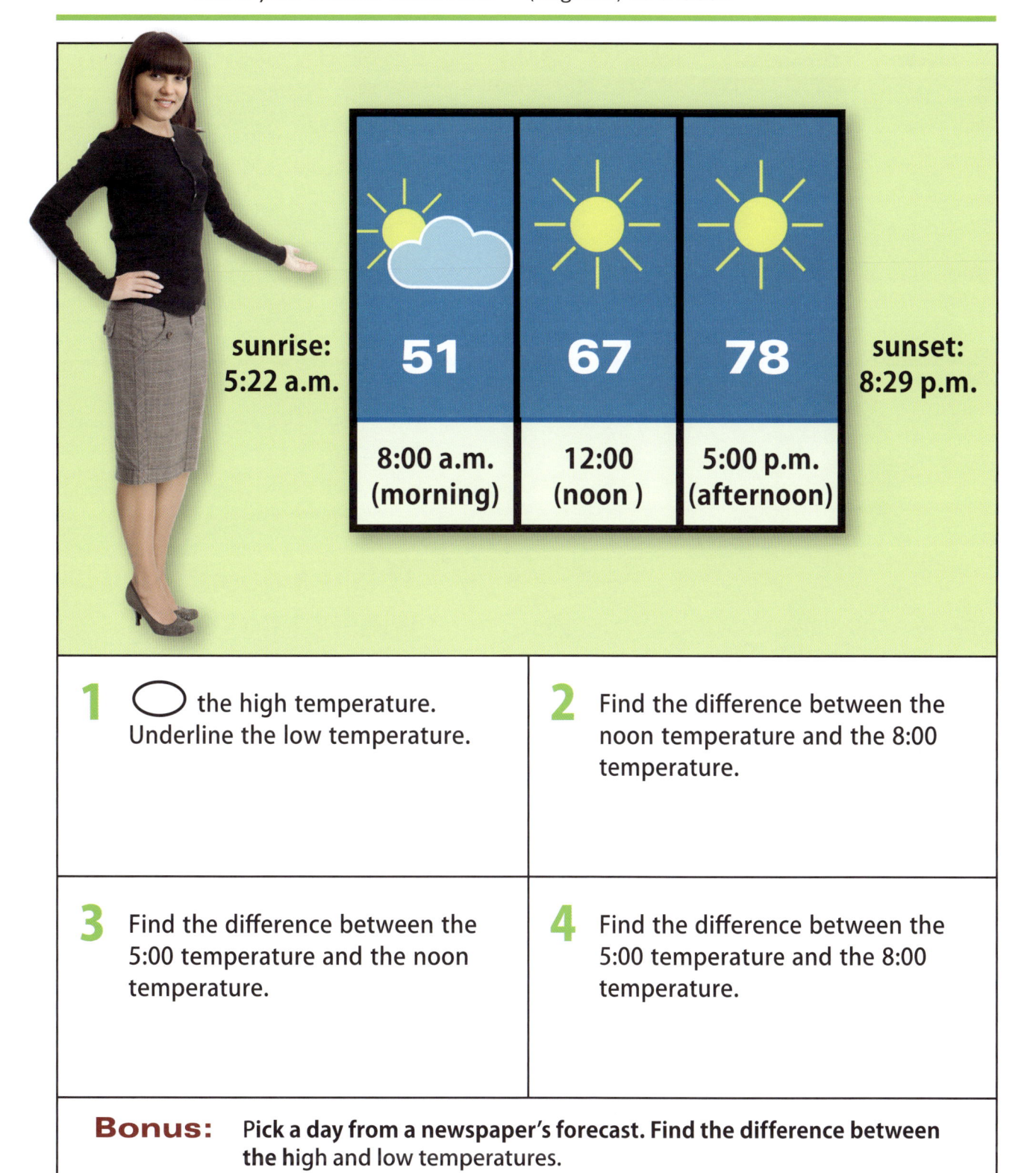

sunrise: 5:22 a.m.

51	67	78
8:00 a.m. (morning)	12:00 (noon)	5:00 p.m. (afternoon)

sunset: 8:29 p.m.

1 ◯ the high temperature. Underline the low temperature.

2 Find the difference between the noon temperature and the 8:00 temperature.

3 Find the difference between the 5:00 temperature and the noon temperature.

4 Find the difference between the 5:00 temperature and the 8:00 temperature.

Bonus: Pick a day from a newspaper's forecast. Find the difference between the high and low temperatures.

Chapter 4 • 0 — 100

A Summer Forecast 1

Date _____

Directions: Tamara is giving a 7-day weather forecast. Use the information below to solve the problems on the next page.

Tuesday	Wednesday	Thursday	Friday	Saturday	Sunday	Monday
High: 100°	High: 98°	High: 87°	High: 83°	High: 97°	High: 87°	High: 99°
Low: 90°	Low: 82°	Low: 75°	Low: 70°	Low: 84°	Low: 72°	Low: 83°

This is a good week to go to the beach!

Bonus: ◯ the day that usually starts a week.

Chapter 4 • 0 — 100

A Summer Forecast 2

Date _____

Directions: Use the 7-day forecast to solve the problems below.
Find the differences between the temperatures.
Hint: The answers must be labeled in ° (degrees).

Problem:	**Show work:**
1 Tuesday high: _____ low: _____	
2 Monday high: _____ low: _____	
3 Wednesday and Friday Wednesday's high: _____ Friday's high: _____	
4 Saturday and Friday. Saturday's low: _____ Friday's low: _____	
5 ◯ the day with the warmest high temperature. Put a ✔ on the day with the lowest temperature.	

Chapter 4 • 0 — 100

A Shopping Trip

Date _____

Directions: Solve to find our how much money Tasha has left. ◯ the money spent. Write the problem below.

Hint: Remember the label must have **$** and a **.** (decimal point).

1

2

3

4

$52.00

Solve:

DVDs on Sale

Date _____

Directions: Solve to find out how much money Hank had left.
Hint: Remember the label.

1

2 20 10 5 1 1 1

3

4

Show work:

Chapter 4 • 0 — 100

109

All in a Day's Work

Date _____

Directions: Write the digital time on the clock to show when Mr. Rodriguez leaves his work.

1 Mr. Rodriguez punches in at 7:00 every day he works.

2 He punches out 8 hours later. What time does he punch out?

Solve:

Mr. Rodriguez earns $15.00 an hour. ◯ the amount of money he earns in a day.

Chapter 4 • 0 — 100

Mr. Rodriguez's Morning

Date _____

Directions: Mr. Rodriguez has a very busy morning work schedule. Draw hands on the clock to match the time.

1 Mr. Rodriguez comes to work at 7:00.

2 He finishes stacking all of the boxes for today's shipment 30 minutes later.

3 1 hour later, he labels all of the boxes to ship for the day.

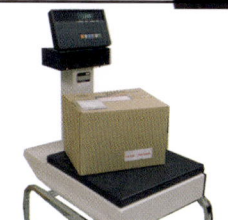

4 It takes 30 more minutes to stack all of the boxes on dollies to put on a truck.

5 30 minutes later, Mr. Rodriguez has finished loading the boxes to ship.

6 It takes him 2 more hours to label and stack all of the boxes for the afternoon shipment.

7 Mr. Rodriguez goes to lunch 30 minutes after he has stacked all of the boxes. When does he go to lunch?

Chapter 4 • 0 — 100

15 Minutes Later

Date _____

Directions: Listen to your teacher read each sentence.
Draw hands on the clock to match the digital time.

1 Martha leaves to walk to school at 7:30.

2 She meets her friend Maria at 7:35.

3 It takes them 10 more minutes to walk to school. They get there at 7:45.

4 The bell rings to enter the building at 8:00.

5 15 minutes later, Martha is sitting in her first class. It is 8:15.

6 It takes Martha 15 minutes to finish a short quiz. It is 8:30.

Bonus:

What time will it be in 15 more minutes? _____

Chapter 4 • 0 — 100

15 Minutes Earlier

Date _____

Directions: Draw hands on the clocks to show what time it was 15 minutes earlier.

1 Ariana finished writing her quiz at 11:00.

2 Ariana started the quiz 15 minutes earlier. At what time did she start?

3 Mrs. Brown finished correcting today's quiz at 4:30.

4 She started to correct the quiz 15 minutes earlier. At what time did she start?

Chapter 4 • 0 — 100

A School Football Game

Date _____

Directions: ◯ the coins that match the total amount.

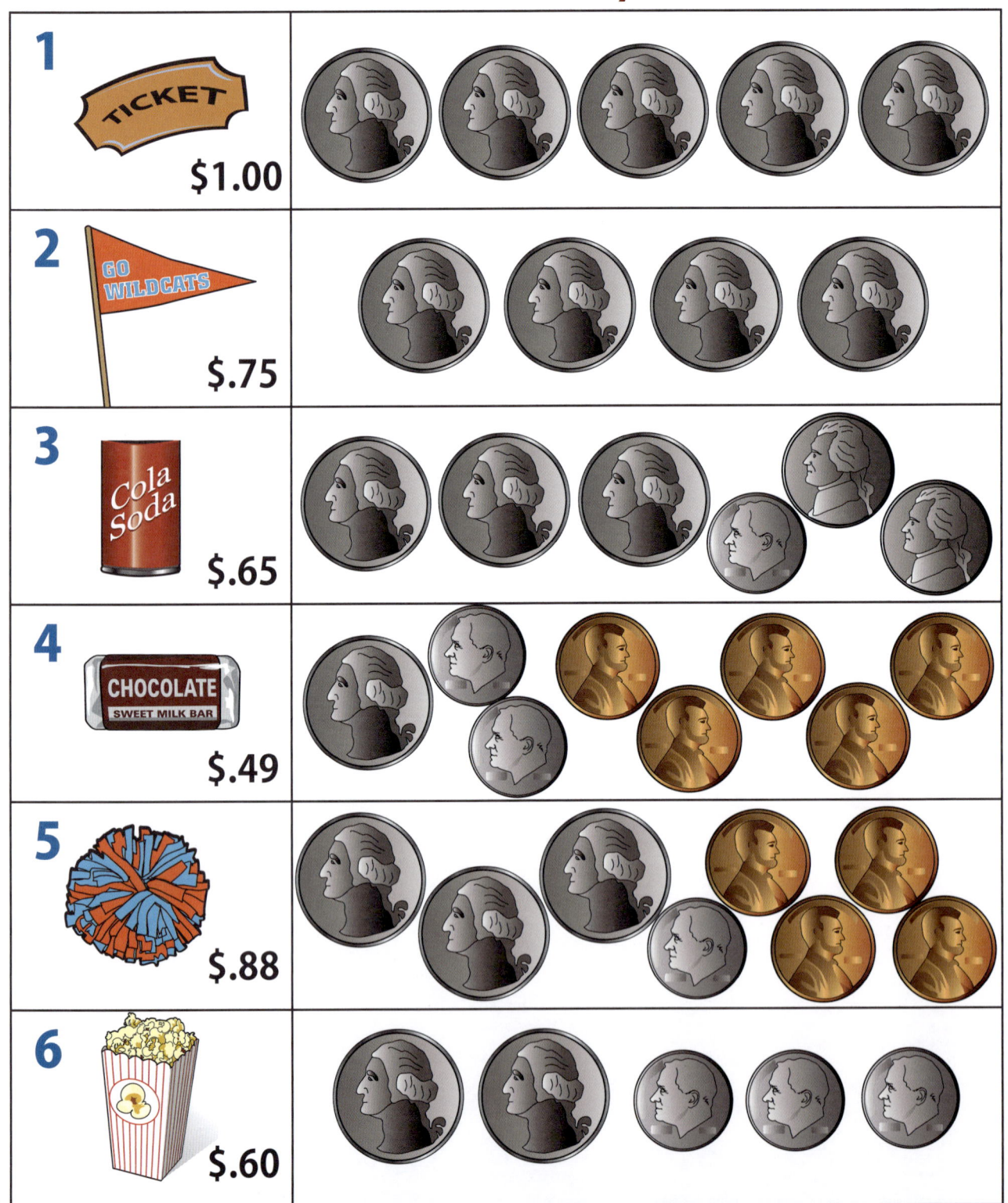

Purchase:	Coins you have:
1. Ticket — $1.00	5 quarters
2. Go Wildcats flag — $.75	4 quarters
3. Cola Soda — $.65	3 quarters, 1 nickel, 1 nickel
4. Chocolate Sweet Milk Bar — $.49	1 quarter, 2 nickels, 5 pennies
5. Pom-pom — $.88	3 quarters, 1 nickel, 4 pennies
6. Popcorn — $.60	2 quarters, 3 nickels

Chapter 4 • 0 — 100

Grocery Store

Date _____

Directions: Kelsey loves bargains. Each week she looks at the sale cart to buy something on sale. ◯ the coins she gets back from each purchase.

	Kelsey has:	Kelsey buys:	Kelsey's change:
Week 1	$1 bill	picture frame $.95	5 pennies, 1 nickel
Week 2	4 quarters	nail polish $.75	3 nickels, 1 dime
Week 3	4 quarters	chocolates $.59	2 pennies, 1 nickel, 1 dime, 1 quarter
Week 4	$1 bill	book $.80	4 nickels, 1 dime

▼ Note: Try to use the least number of coins.

Chapter 4 • 0 — 100

115

A Night Out

Date _____

Directions: Katrina is going out with friends to eat and see a movie. X (cross out) the coins and bills that equal the total amount she pays each time she makes a purchase.

Purchases:

1 $1.00 Hint: Use coins.

2 $4.00

3 $10.00

4 $5.00

5 $20.00

6 Create your own.

Katrina's cash

Bonus: ◯ the money that is left and tell the amount of money Katrina now has.

Chapter 4 • 0 — 100

Dining Out

Date _____

Directions: Help the server make change. ◯ the fewest coins/bills each customer would get back after paying for a meal.

Customer gave:	Customer ordered:	Change back:
1 — three $10 bills	Total = $24.00	two quarters, $5, $1, $1, $10
2 — $5 and $10	Total = $14.10	nickel, dime, dime, quarter, quarter, quarter, $1, $1, $1, $1
3 — $20	Total = $17.24	penny, penny, quarter, quarter, $1, $1, $1, $1
4 — $10	Total = $6.00	 $1, $1, $1, $1, $1, $1
5 — three $20 bills	Total = $52.00	$1, $1, $1, $1, $5, $5, $10

Bonus: Create your own money problem for someone to solve.

Chapter 4 • 0 — 100

117

A Line Graph 1

Date _____

Directions: Shawn is saving his tips as a server at a local restaurant to buy a new "smart" cell phone. His goal is to save $100.000. Use the graph below to answer the questions on the next page.

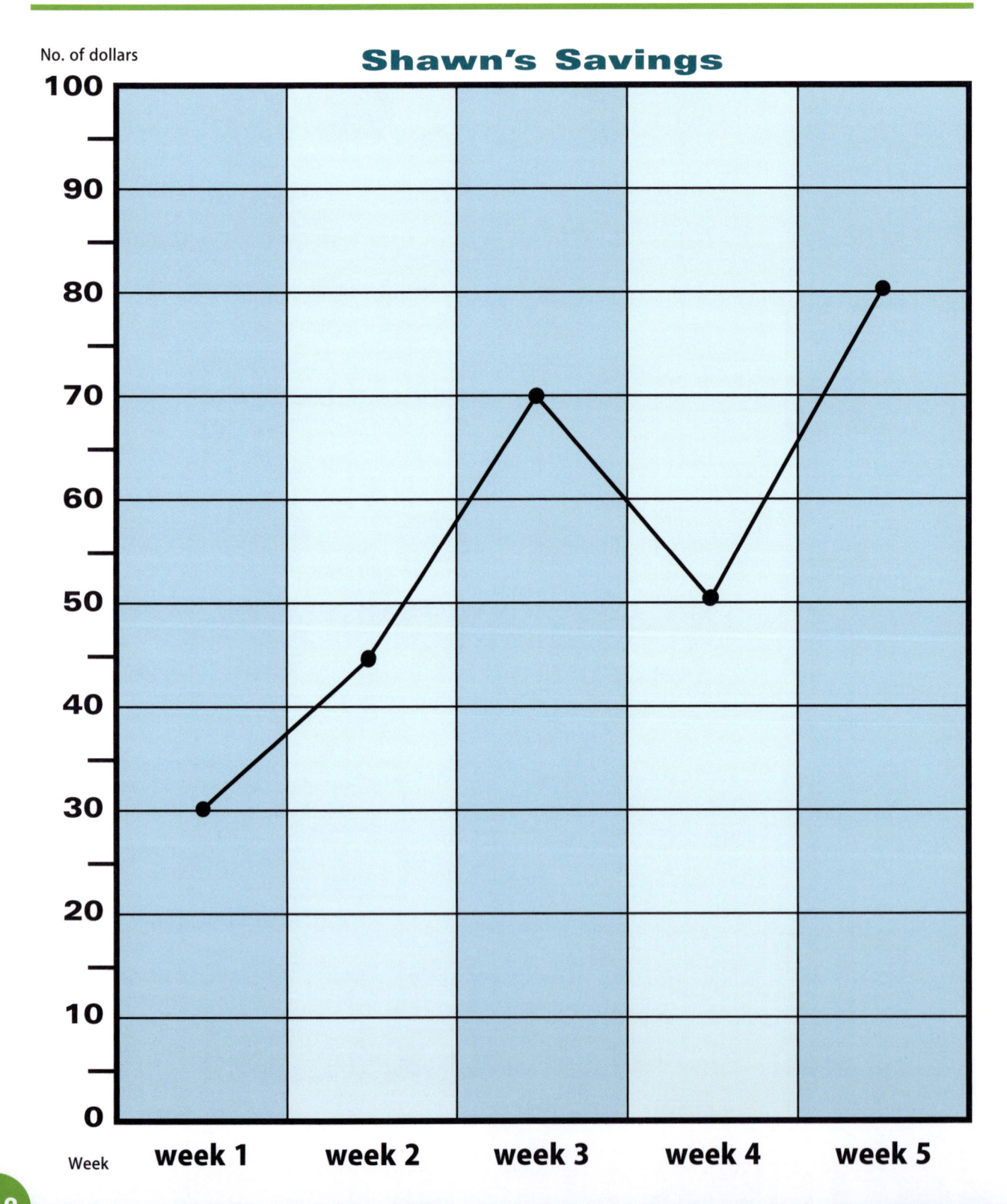

Shawn's Savings

No. of dollars

Week	week 1	week 2	week 3	week 4	week 5
Dollars	30	45	70	50	80

Chapter 4 • 0 — 100

A Line Graph 2

Date _____

Directions: Listen to your teacher read the problems.
Show work and remember to label your answers.

1 How much money did Shawn earn the first week?	**2** How much more money did he earn in week 2?
3 How much more money did Shawn earn in week 3 than week 2?	**4** In week 4, Shawn had to take some money out of his savings account to help pay for an oil change for his car. How much money does he have left?
5 How much money did Shawn put into his account in week 5?	**6** How much more money does Shawn need to make before he gets to $100.00?

Chapter 4 • 0 — 100

Morning Sales 1

Date _____

Directions: Miguel owns a newsstand. Graph how many newspapers and magazines he sold in one morning.

Miguel's Tally

1.	The Daily News	82
2.	The Morning Star	76
3.	City Weekly	55
4.	Popular Sports	35
5.	Fashion Magazine	60
6.	Wildlife Photography	23

Bonus: ◯ the newspaper or magazine that had the lowest sales.

Chapter 4 • 0 — 100

Morning Sales 2

Date _____

Directions: Graph the newspapers and magazines that Miguel sold.

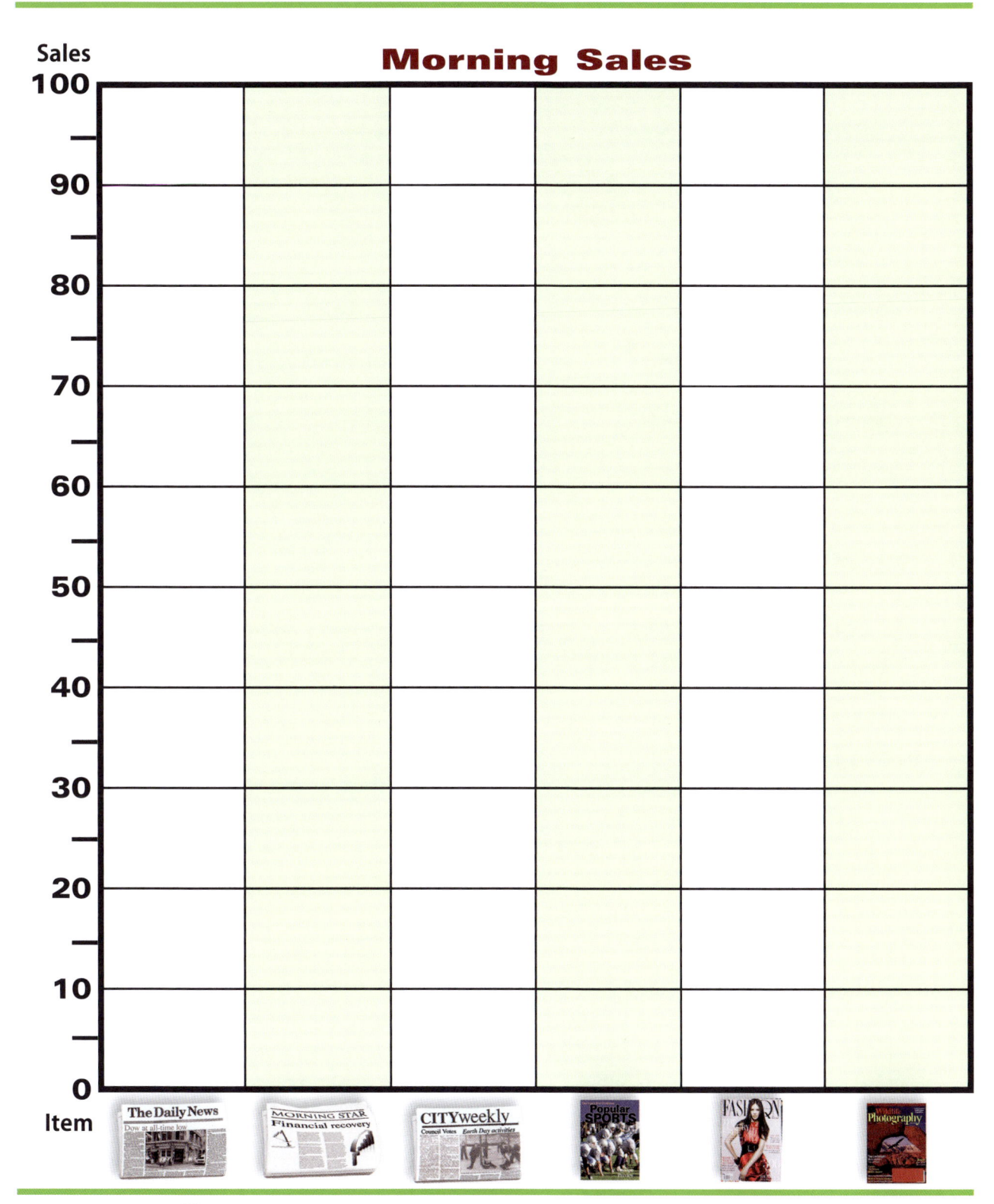

Morning Sales

Sales: 100, 90, 80, 70, 60, 50, 40, 30, 20, 10, 0

Item: The Daily News, Morning Star, CITYweekly, Popular Sports, Fashion, Photography

Bonus: ◯ the column that shows the greatest amount sold.

Chapter 4 • 0 — 100

Game Points 1

Date _____

Directions: Dominick and Shawna are the top scorers for their basketball team. Pick one of the players to graph on the next page.

Dominick's Points

Game 1   24 points
Game 2   42 points
Game 3   30 points
Game 4   16 points
Game 5   38 points
Game 6   26 points

Shawna's Points

Game 1   18 points
Game 2   46 points
Game 3   24 points
Game 4   20 points
Game 5   32 points
Game 6   40 points

Challenge:

1. How many points has Dominick made so far this season?

   _____ points

2. Dominick wants to make 200 points before the season is over. How many more points must he make to reach his goal?

   _____ points

1. How many points has Shawna made so far this season?

   _____ points

2. Shawna wants to make 200 points before the season is over. How many more points must she make to reach her goal?

   _____ points

Game Points 2

Date _____

Directions: Using the charts on the previous page, pick either Dominick or Shawna and graph the points made during this season's games.

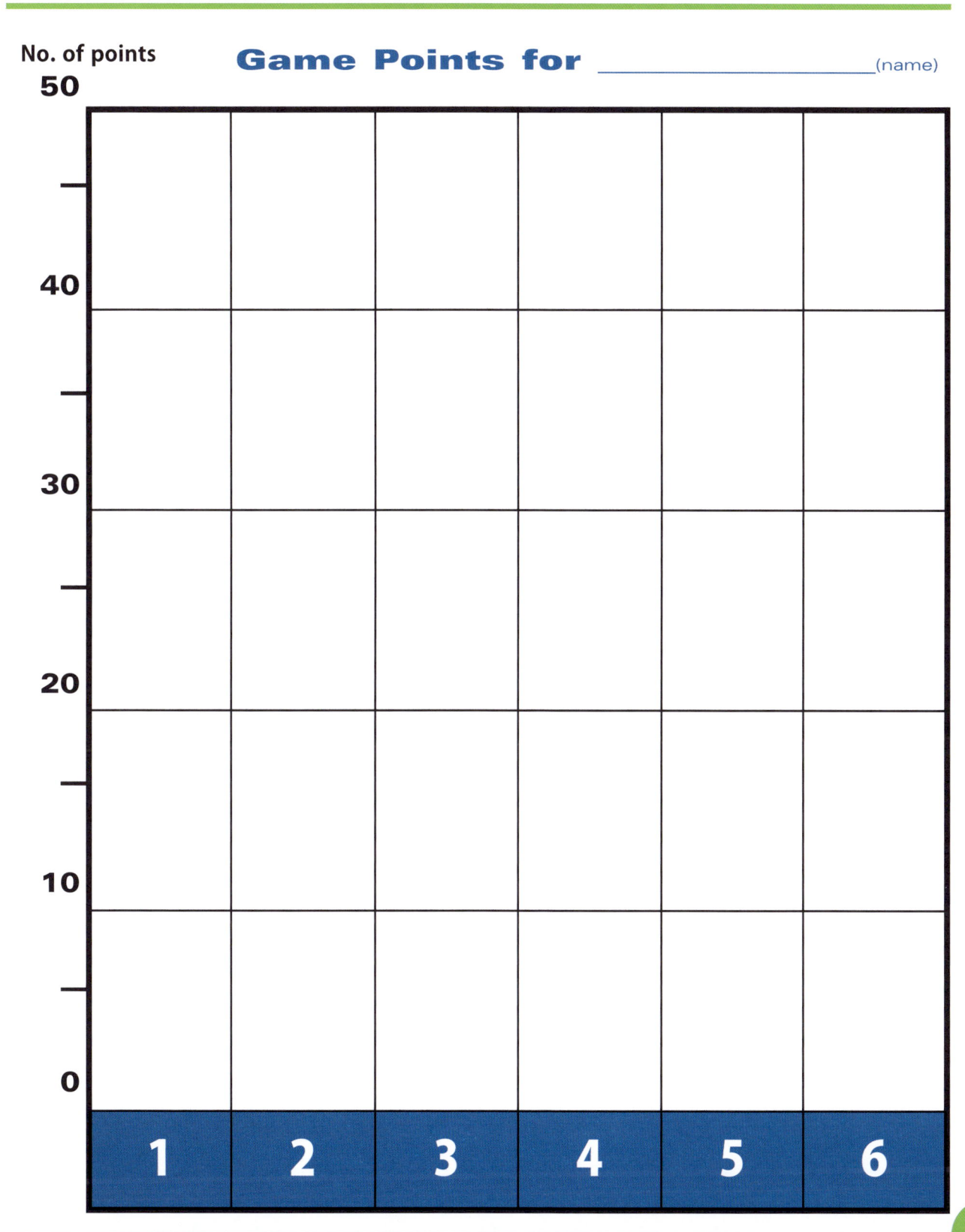

Game Points for _____ (name)

No. of points

50
40
30
20
10
0

1 2 3 4 5 6

Chapter 4 • 0 — 100

123

Number Riddles

Date _____

Directions: Use the 100s Chart (p. 90) to answer the riddles below.

1

I am thinking of a number between 14 and 26.

The number is 6 more than 16. The number I am thinking about is:

_____

2

I am thinking of a number between 40 and 50.

The number is 6 more than 58 - 17. The number I am thinking about is:

_____

3

I am thinking of a number between 70 and 80.

The number is 4 more than 62 + 8. The number I am thinking about is:

_____

Finding Petrovich

Date _____

Directions: Listen to your teacher read the clues. Write the initial (or beginning letter) that matches the person who is standing in line. ◯ Petrovich.

Clues

1. Put an **M** on Maggie, who is **1st in line**.
2. Put a **J** on Juan, who is standing **last in line**.
3. Put an **A** on Ariana, who is **5th in line**.
4. Put a **D** on Dominick, who is standing **behind Ariana**.
5. Put an **S** on Scott, who is **in front of Ariana**.
6. Put a **T** on Tom, who is **3rd in line**.
7. Put an **N** on Naomi, who is standing **between Maggie and Tom**.
8. Put an **H** on Helen, who standing **in front of Juan**.
9. Put a **B** on Barbara, who is standing **behind Petrovich**.
10. ◯ **Petrovich**.

Bonus: Create your own problem.

Chapter 4 • 0 — 100

125

Chapter 5
0 — 1000

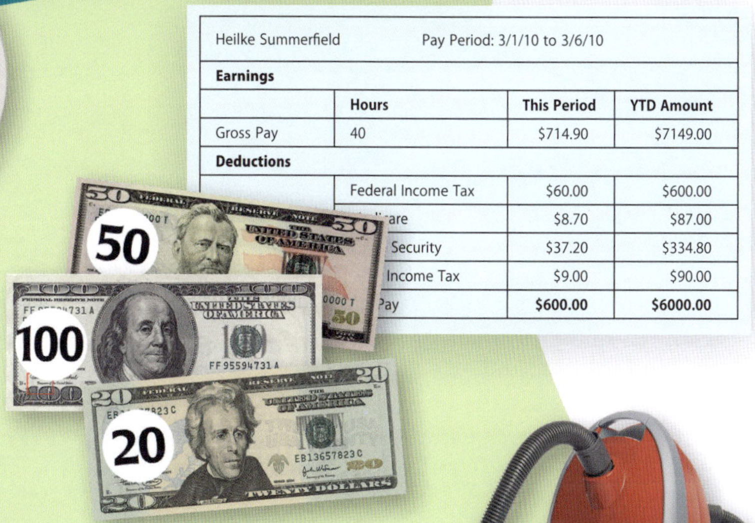

Heilke Summerfield	Pay Period: 3/1/10 to 3/6/10		
Earnings			
	Hours	This Period	YTD Amount
Gross Pay	40	$714.90	$7149.00
Deductions			
	Federal Income Tax	$60.00	$600.00
	Medicare	$8.70	$87.00
	Social Security	$37.20	$334.80
	State Income Tax	$9.00	$90.00
	Net Pay	$600.00	$6000.00

Chapter 5 • 0 — 1000

A New Bike

Date _____

Directions: Willie bought a new bike. ◯ the money that matches the total price.

$375.00

Bonus: If Willie gave the clerk $400.00 to buy the bike, how much would he get back? _____

128

Chapter 5 • 0 — 1000

Car Repair

Date _____

Directions: Nancy had her car repaired and bought new tires.
◯ the total amount of money she needs to pay.

Nevin's Garage

Repair		Total
Change Oil		$36.00
Tires		$248.00
Brakes		$250.00
Wipers		$44.00
	Total	$578.00

Bonus: How much money does Nancy have left? _____

Chapter 5 • 0 — 1000

A Trip Around the State 1

Date _____

Directions: Use the map to solve the problems on the next page.

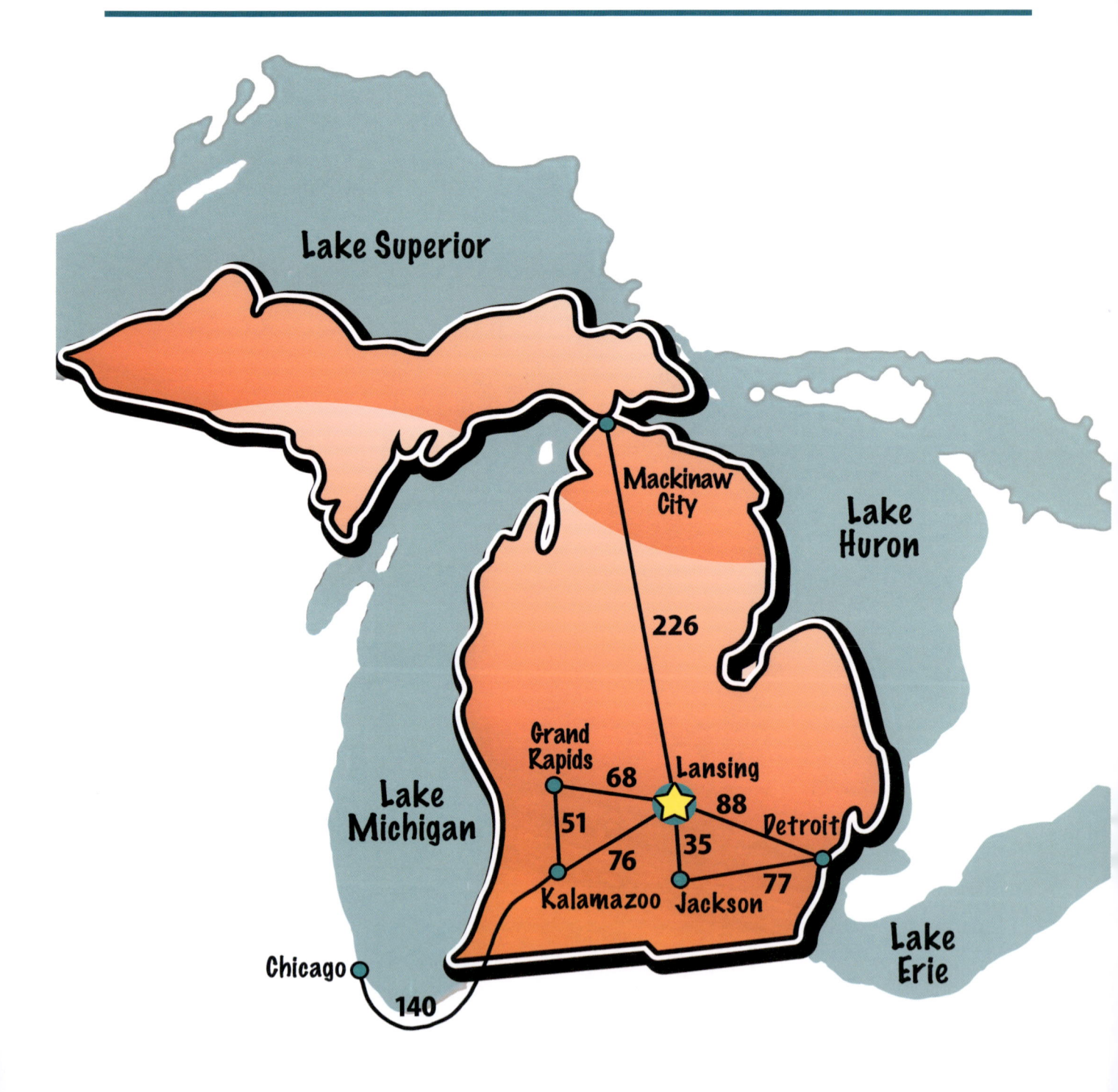

- Lake Superior
- Lake Huron
- Lake Michigan
- Lake Erie
- Mackinaw City
- Grand Rapids
- Lansing ★
- Detroit
- Kalamazoo
- Jackson
- Chicago

Distances:
- Mackinaw City to Lansing: 226
- Grand Rapids to Lansing: 68
- Lansing to Detroit: 88
- Grand Rapids to Kalamazoo: 51
- Kalamazoo to Lansing/Jackson: 76
- Lansing to Jackson: 35
- Jackson to Detroit: 77
- Chicago to Kalamazoo: 140

Bonus: The ★ shows the capital city. ◯ the capital city on the map.

130   Chapter 5 • 0 — 1000

A Trip Around the State 2

Date _____

Directions: Answer the questions below. **Hint:** Show work and remember labels!

1 Put an **X** on Lake Superior. Underline the words "Lake Huron." Write an **M** on Lake Michigan.

2 The Johnson family lives in Detroit. They are going to Grand Rapids. They will drive through Lansing. How many miles will they drive altogether?

3 The Gomez family lives in Lansing and are taking a trip to Chicago. How many miles will they drive altogether?

4 The Lang family lives in Mackinaw City and are taking a trip to Jackson. How many miles will they drive altogether?

Bonus: Create and share your own problem using the map.

Capacity Problems

Date _____

Directions: Use the chart below to solve the problems.

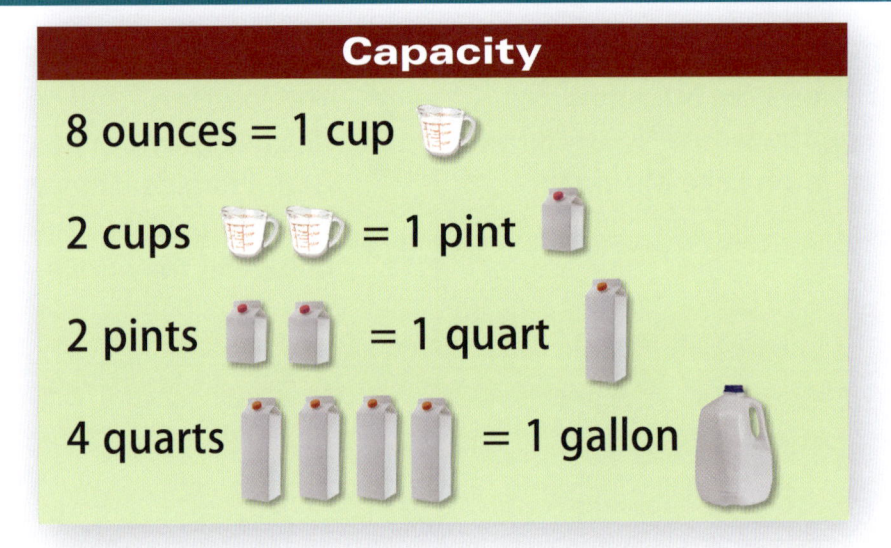

Capacity

8 ounces = 1 cup

2 cups = 1 pint

2 pints = 1 quart

4 quarts = 1 gallon

Problem: | **Show Work:**

1 The cat knocked over 3 cups of milk. How many ounces were knocked over?

2 Aunt Martha canned 5 quarts of tomatoes. How many pints did she can?

3 Henry bought 4 gallons of ice cream. How many quarts of ice cream did he buy?

Bonus: Tell how many pints equals one gallon of ice cream. _____

132 Chapter 5 • 0 — 1000

Weight Lifting

Date _____

Directions: DeAndre lift weights at the local gym.
Solve to find out how many pounds he can lift.
Hint: Remember that your answer must have pounds (**lbs.**) as a label.

125 lbs.   125 lbs.

Show work:

Bonus: A gallon of water weighs 8 pounds.
How many pounds would 3 gallons of water weigh? _____

Chapter 5 • 0 — 1000

NFL Players 1

Date _____

Directions: Read the weight of each team member and use the weights to solve the problems on the next page.

- Player 6: 235
- Player 29: 291
- Player 35: 320
- Player 42: 244
- Player 18: 312

Bonus: ◯ the player who weighs the most. Put an **X** on the player who weighs the least.

NFL Players 2

Date _____

Directions: Use the weights of the players to solve the problems below.

1 Players 6 and 18 tackled a member of the opposite team. How many pounds do the two players weigh altogether?

2 Players 35 and 42 made the next tackle. How many pounds do these players weigh altogether?

3 Players 29 and 35 made a tackle. How much do these two players weigh?

4 Add the weights of the 2 players who weigh the **most.**

5 Add the weights of the 2 players who weigh the **least.**

Bonus: Add the weight of players 29, 42, and 18.

A Monthly Budget 1

Date _____

Directions: Heilke works part-time after school and on weekends as a shipping clerk. She takes home $364.00 a month. Below are her expenses this month.

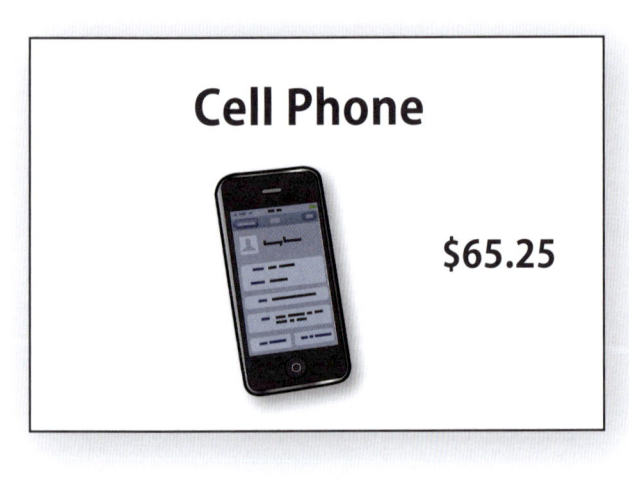

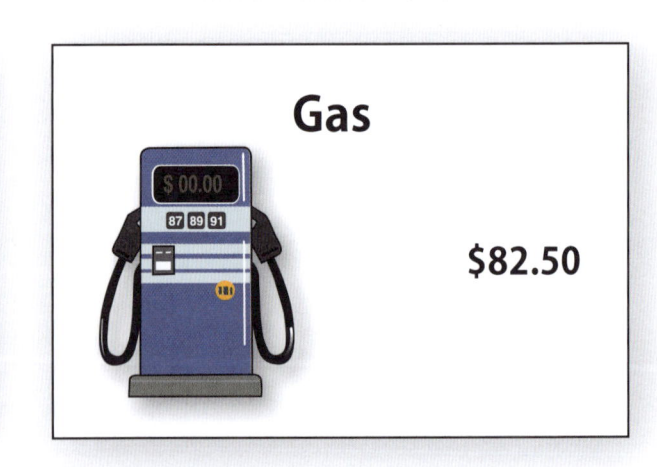

Car Payment	Savings
$156.00	$20.00

Cell Phone	Gas
$65.25	$82.50

Entertainment	Downloads for iPod
$28.25	$12.00

Bonus: How much money does Heilke have left after she has budgeted her money? _____

Chapter 5 • 0 — 1000

A Monthly Budget 2

Date _____

Directions: Use Heilke's Monthly Budget (page 136) to solve the problems below.
Hint: Remember to use the **$** and **.** in your answer.

Problem:	Show work:
1 Heilke must write checks for (cell phone) (car payment) How much money did she spend?	
2 How much money did Heilke spend on her car this month? (car payment) (gas)	
3 How much money did Heilke spend on (entertainment) (iPod)	
4 Tell one reason why Heilke puts money into a savings account. _____ How much does she put into her account? _____	
Bonus: How much more money must Heilke save to buy tennis shoes that cost $90.00? _____	

Chapter 5 • 0 — 1000

Food Drive

Date _____

Directions: The 9th grade has collected 750 cans for a food drive. How many more cans do the students need to collect to reach the goal of 1,000 cans of food?

Show work:

Heilke's Savings

Date _____

Directions: Solve the problem below.

1 Heilke had some money in her savings account.

2 She got this much money for her birthday:

50 20
10 5

3 Now Heilke has this amount:

100 100
50   20
10   10
5

4 ◯ the money Heilke had before she got her birthday money.

100 100
50   20
10   10
5

Show work:

Bonus: How much money will Heilke have left if she buys tennis shoes for $90.00? _____

Chapter 5 • 0 — 1000

139

Heilke's Paycheck 1

Date _____

Directions: Use Heilke's paycheck to answer the questions on the next page.

Ever-Ready Shipping
5678 B Street
Anytown, WI 53705

Check Number: **6640**
Date: 3/1/10

PAY: === **Three hundred sixty-four and no cents** ========== **$364.00**

Pay to the order of: Heilke Summerfield
                         1098 Elm Avenue
                         Anytown, WI 53705

Heilke Summerfield   Pay Period: 2/1/10 to 2/28/10

Earnings			
	Hours	This Period	YTD Amount
Gross Pay	40	$408.67	$1226.01
Deductions			
	Federal Income Tax	$00.00	$00.00
	Medicare	$5.28	$15.84
	Social Security	$30.65	$91.95
	State Income Tax	$00.00	$00.00
	Net Pay	**$372.74**	**$1118.22**

▼ Note: Heilke does not make enough on her part-time job to pay either federal or state income tax.

Chapter 5 • 0 — 1000

Heilke's Paycheck 2

Date _____

Directions: Use Heilke's paycheck and paycheck stub (page 140) to answer the questions below.

---

1 ◯ the dates that Heilke worked to get her paycheck.

2 On a paycheck stub, the gross amount means how much money a person earns before any taxes are taken out. Write the gross amount Heilke earned on the line.

Gross pay _____

3 Everyone must pay Medicare tax no matter how much they earn. Write the amount of Medicare tax Heilke paid for this period.

Medicare _____

4 Everyone must pay Social Security tax no matter how much they earn. Write the amount of Social Security tax Heilke paid for this period.

Social Security _____

5 Net Pay means the amount of money Heilke takes home after all of the taxes are taken out of her check. Underline the amount on her paycheck.

6 Find the difference between Heilke's gross pay and net pay to find out how much she had to pay in taxes.

_____

Challenge: YTD means the amount earned so far this year. Find the difference between Heilke's gross YTD amount and her net YTD amount.

_____

Mr. Sampson Gets Paid 1

Date _____

Directions: Mr. Sampson is Heilke's boss. He gets paid each week. Below is a copy of his paycheck. Use it to solve the problems on the next page.

Ever-Ready Shipping
5678 B Street
Anytown, WI 53705

Check Number: **6702**
Date: 3/12/10

PAY: === **Six hundred and no cents** =================== **$600.00**

Pay to the order of: Melvin Sampson
1112 Maple Street
Anytown, WI. 53704

Heilke Summerfield   Pay Period: 3/1/10 to 3/6/10

Earnings

	Hours	This Period	YTD Amount
Gross Pay	40	$714.90	$7149.00

Deductions

Federal Income Tax	$60.00	$600.00
Medicare	$8.70	$87.00
Social Security	$53.62	$536.20
State Income Tax	$9.00	$90.00
Net Pay	**$583.58**	**$5835.80**

Challenge: What is Mr. Sampson's YTD Net Pay? _____

Chapter 5 • 0 — 1000

Mr. Sampson Gets Paid 2

Date _____

Directions: Use Mr. Sampson's paycheck and paycheck stub (page 142) to answer the questions below.

1 ◯ the dates that Mr. Sampson worked to get his paycheck.

2 Underline the net amount or the amount of money Mr. Sampson takes home.

3 How many hours did Mr. Sampson work in the week?

_____ hours

4 Write the amount of Federal income tax Mr. Sampson paid this week.

Federal Income Tax _____

5 Write the amount of Medicare tax Mr. Sampson paid this week.

Medicare _____

6 Write the amount of social security tax Mr. Sampson paid this week.

Social Security _____

7 Write the amount of state income tax Mr. Sampson paid this week.

State Income Tax _____

8 Find the difference between Mr. Sampson's gross pay and net pay this period to find out how much he had to pay in taxes.

_____ Gross Pay
− _____ Net Pay

Chapter 5 • 0 — 1000

A Saturday Morning

Date _____

Directions: Draw hands on each clock to match the time.

	Time:	**Clock:**
1	Shaun wakes up at 7:00.	
2	5 minutes later he is brushing his teeth.	
3	15 minutes later he eats breakfast.	
4	20 minutes later he packs his gym bag to go to soccer practice.	
5	He waits 10 minutes for his ride to practice.	
6	25 minutes later he gets to practice. What time is it?	

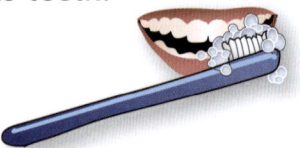

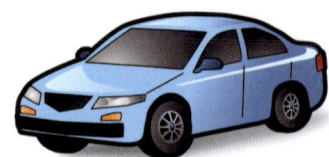

Chapter 5 • 0 — 1000

How Many Minutes?

Date _____

Directions: Draw hands on the analog clocks to show the correct time.

1a Henrietta finished her math homework at 7:45.

1b It took Henrietta 32 minutes to do her homework. What time **did** she start?

2a Rosa finished text messaging her friends at 8:30.

2b She started 4 minutes ago. What time did she start?

Chapter 5 • 0 — 1000

A Day's Work

Date _____

Directions: Write the digital times on the time clock.
Solve to find out how much money Ms. DiMartino makes in one day.

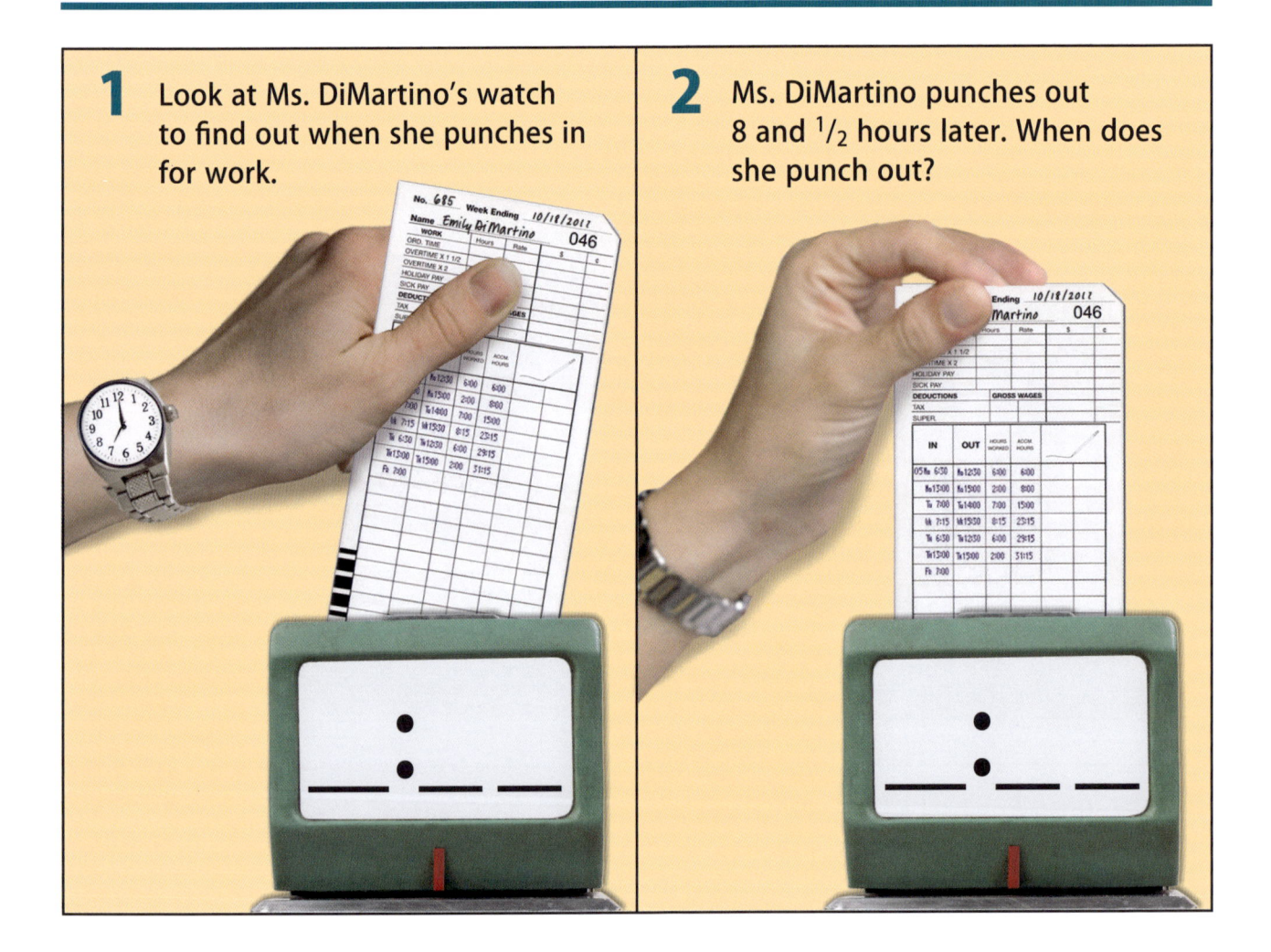

1 Look at Ms. DiMartino's watch to find out when she punches in for work.

2 Ms. DiMartino punches out 8 and $^1/_2$ hours later. When does she punch out?

___ : ___

___ : ___

Directions: ◯ the total amount of money that Ms. DiMartino makes in one day if she gets $10.00 an hour.

▼ Note: Ms. DiMartino does not get paid for taking a 30-minute lunch.

Chapter 5 • 0 — 1000

A Week's Work

Date _____

**Directions: Solve to find out how much each person gets paid weekly.
Write the total amount on the blanks.
Hint:** Most workweeks are 5 days.

1 Mr. Brooks gets paid $50.00 a day. Solve to find out how much he gets paid in a week.

Sunday	Monday	Tuesday	Wednesday	Thursday	Friday	Saturday
off	50	50	50	50	50	off

Mr. Brook's weekly paycheck is: _____

2 Mrs. Gupta gets paid $110.00 a day. Solve to find out how much she gets paid in a week.

Sunday	Monday	Tuesday	Wednesday	Thursday	Friday	Saturday
off	100 + 10	100 + 10	100 + 10	off	100 + 10	100 + 10

Mrs. Gupta's weekly paycheck is: _____

Bonus: Create and share your own problem using a weekly salary.

Chapter 5 • 0 — 1000

Time Zones 1

Date _____

Directions: If all 50 states are included, the United States of America has 6 different times zones. Read the clock for each time zone. Use this map to solve the problems on the next page.

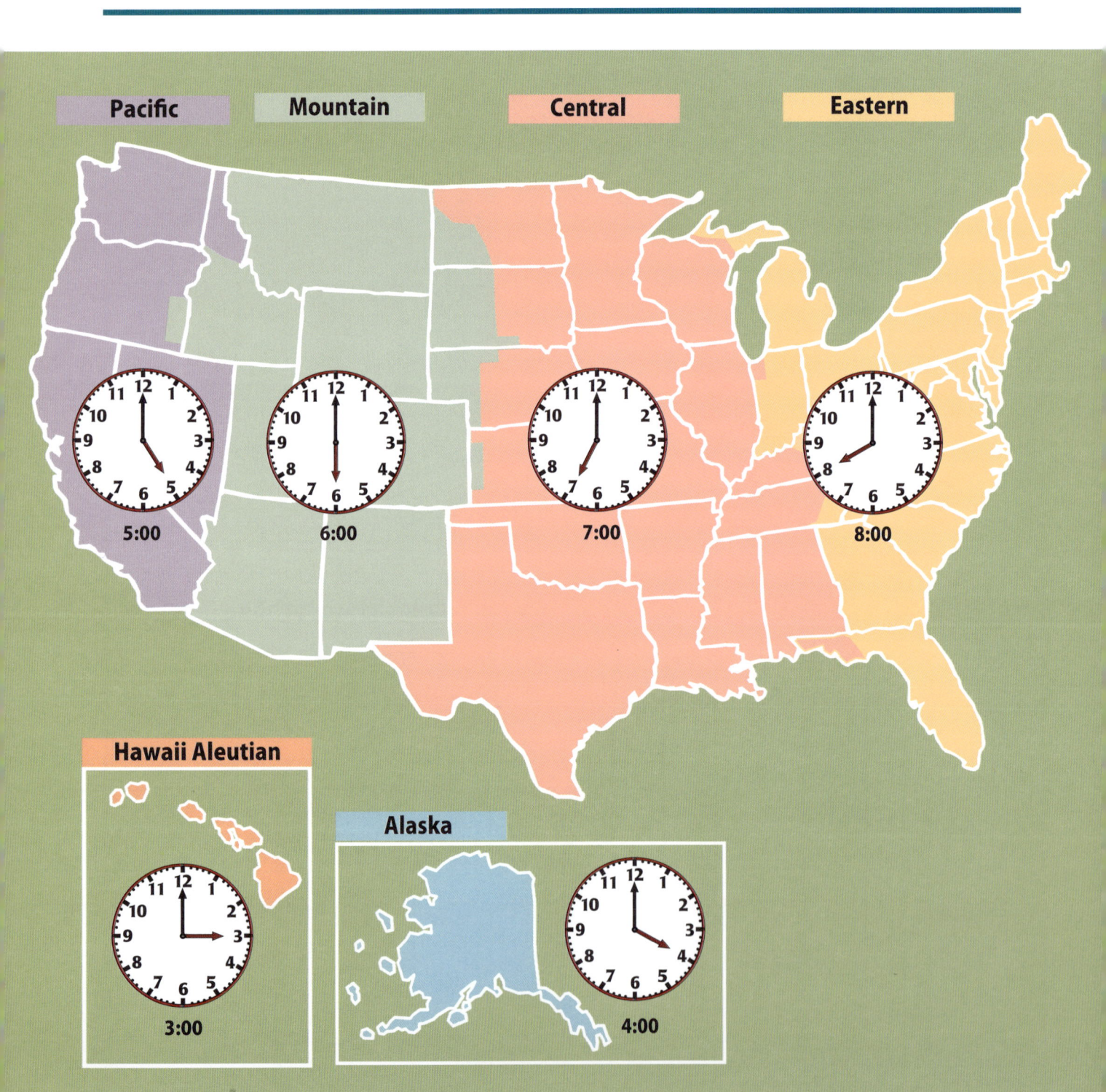

Pacific — 5:00
Mountain — 6:00
Central — 7:00
Eastern — 8:00
Hawaii Aleutian — 3:00
Alaska — 4:00

Chapter 5 • 0 — 1000

Time Zones 2

Date _____

Directions: Use the time zone map on page 148 to help solve the problems below.

1. Joe lives in the Eastern Time Zone and wants to text message his cousin Arnie who lives in the **Pacific Time Zone.** It is **8:00** where Joe lives. Draw hands on the clock to show what time it is where Arnie lives.

2. Mrs. Paulson is on a cruise ship in Alaska. She wants to call her daughter who lives in the **Mountain Time Zone.** It is **4:00** in Alaska. Draw hands on the clock to show what time it is where Mrs. Paulson's daughter lives.

3. Marty is vacationing in Hawaii. He wants to call his friend José who lives in the **Central Time Zone.** It is **3:00** in Hawaii. Draw hands on the clock to show what time it is where José lives.

Chapter 5 • 0 — 1000

A New TV

Date _____

Directions: Mr. Sampson bought a TV. ◯ the amount of money he has left after buying the TV.

Mr. Sampson has:

100, 100, 100, 100
100, 100, 50, 50
50, 20, 20, 5
1, 1, 1, 1

He buys:

special! $798

He has left:

100, 100, 100, 100
100, 100, 50, 50
50, 20, 20, 5
1, 1, 1, 1

Chapter 5 • 0 — 1000

Appliance Sale

Date _____

Directions: ✘ out the money you spend. ◯ the money you have left.

	You have:	You bought:	Money left after purchase:
1	100, 50, 10, 5, 1, 1	 Microwave — $156	100, 50, 10, 5, 1, 1
2	100, 100, 100, 100, 50, 20, 20, 1, 1, 1	Refrigerator — $492	100, 100, 100, 100, 50, 20, 20, 1, 1, 1
3	100, 100, 100, 100, 50, 10, 10, 1, 1, 1	Washer — $463	100, 100, 100, 100, 50, 10, 10, 1, 1, 1
4	100, 100, 50, 10, 10, 5, 1, 1, 1, 1	Air conditioner — $229	100, 100, 50, 10, 10, 5, 1, 1, 1, 1

Chapter 5 • 0 — 1000

151

Big Box Store 1

Date _____

Directions: Mr. Sampson is a manager in a big box store. He keeps track of merchandise that is sold. Use the tally chart below to graph this month's sales.

Small Appliance Sales

Appliance	Sales
irons	450
blenders	125
vacuum cleaners	70
toasters	730
coffeemakers	325

Bonus: ◯ the appliance that sold the most.

Chapter 5 • 0 — 1000

Big Box Store 2

Date _____

Directions: Complete the bar graph using the tally chart on the page 152.

Shawn's Savings

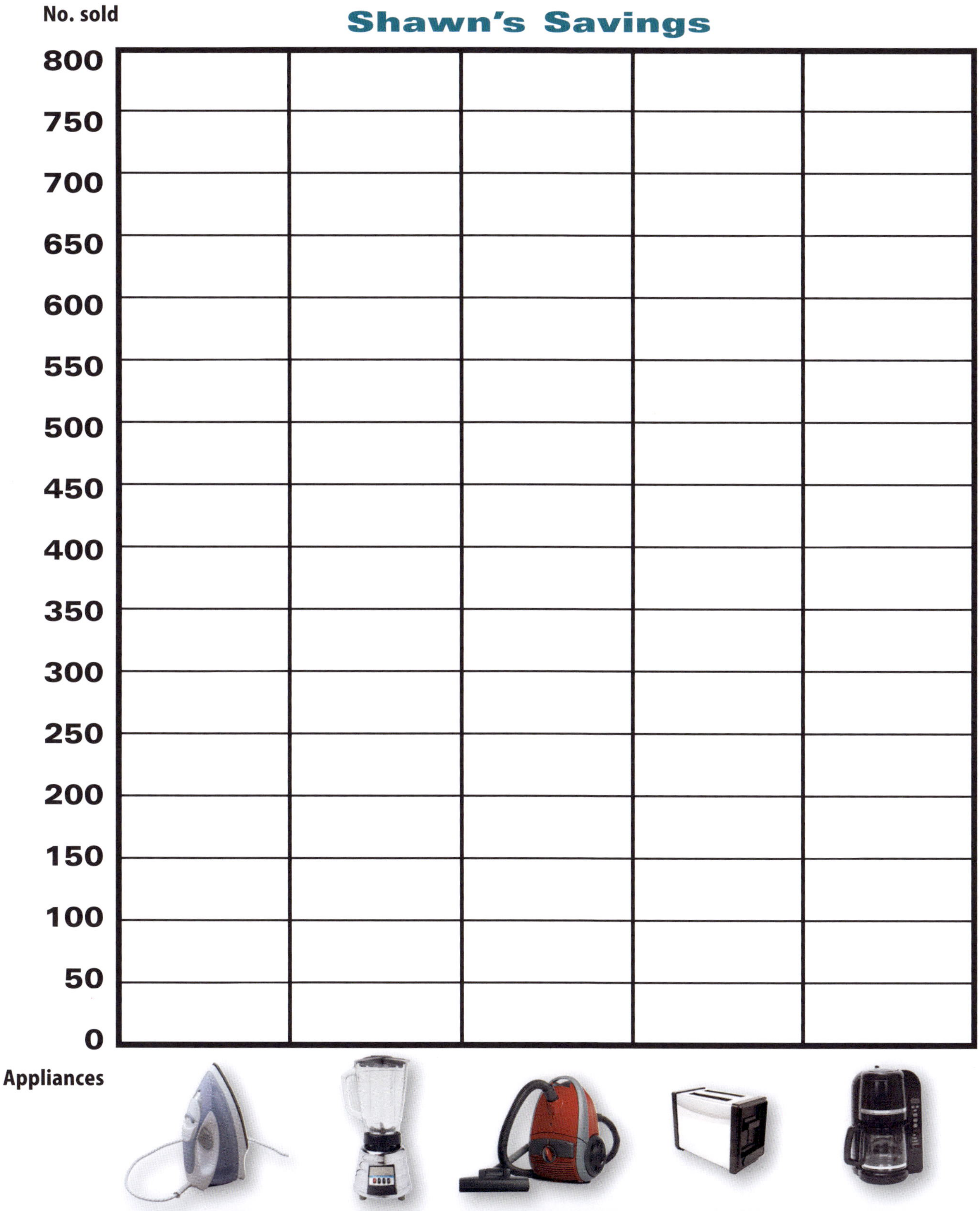

No. sold

| 800 |
| 750 |
| 700 |
| 650 |
| 600 |
| 550 |
| 500 |
| 450 |
| 400 |
| 350 |
| 300 |
| 250 |
| 200 |
| 150 |
| 100 |
| 50 |
| 0 |

Appliances

Chapter 5 • 0 — 1000

153

Student Buses

Date _____

Directions: Write the total number of students in each bus on the bus.

1 Clues
1. Bus B has 41 students in it.
2. Bus A has 4 more students than bus B.
3. Bus C has 5 less students than bus A.
4. Bus D has 8 more students than bus C.

Bus A   **Bus B**   **Bus C**   **Bus D**

2 Clues
1. Bus E has 58 students.
2. Bus G has 6 less students than bus E.
3. Bus F has 4 less students than bus G.
4. Bus H has 7 more students than bus G.

Bus E   **Bus F**   **Bus G**   **Bus H**

Student Parking Lot

Date _____

Directions: Write the first letter of each student's name below the car that belongs to the student.

Clues

1. Shannon's car is between the green and the blue car.
2. Eric's car is next to Shannon's car, but is not green.
3. Nancy's car is not green or yellow.
4. Dominick's car is between the blue and the brown car.
5. What color car belongs to Rosa? _____

green   red   blue   yellow   brown

Chapter 5 • 0 — 1000

Chapter 6
Fractions

157
Chapter 6 • Fractions

Football Snack

Date _____

Directions: Imani and Trevon ordered a pizza to eat while watching a game. They want each person to get the same amount and need to cut the pizza in half. Shade 1/2 of the pizza. Write the fraction 1/2 in the box.

/

Bonus: Write one whole as a fraction.

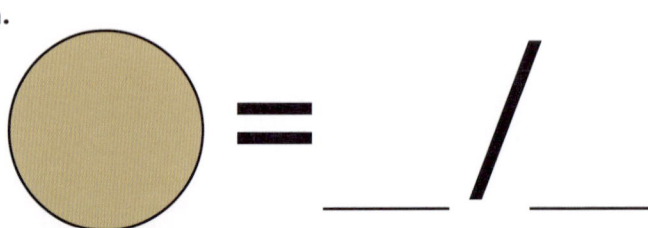

◯ = ___/___

Chapter 6 • Fractions

Game Time

Date _____

Directions: ◯ 1/2 of the team. Write the fraction.

/

Directions: Imani and Trevon want to share the food so that each person gets 1/2 of it. ◯ 1/2 of the food and write the fraction in each box.

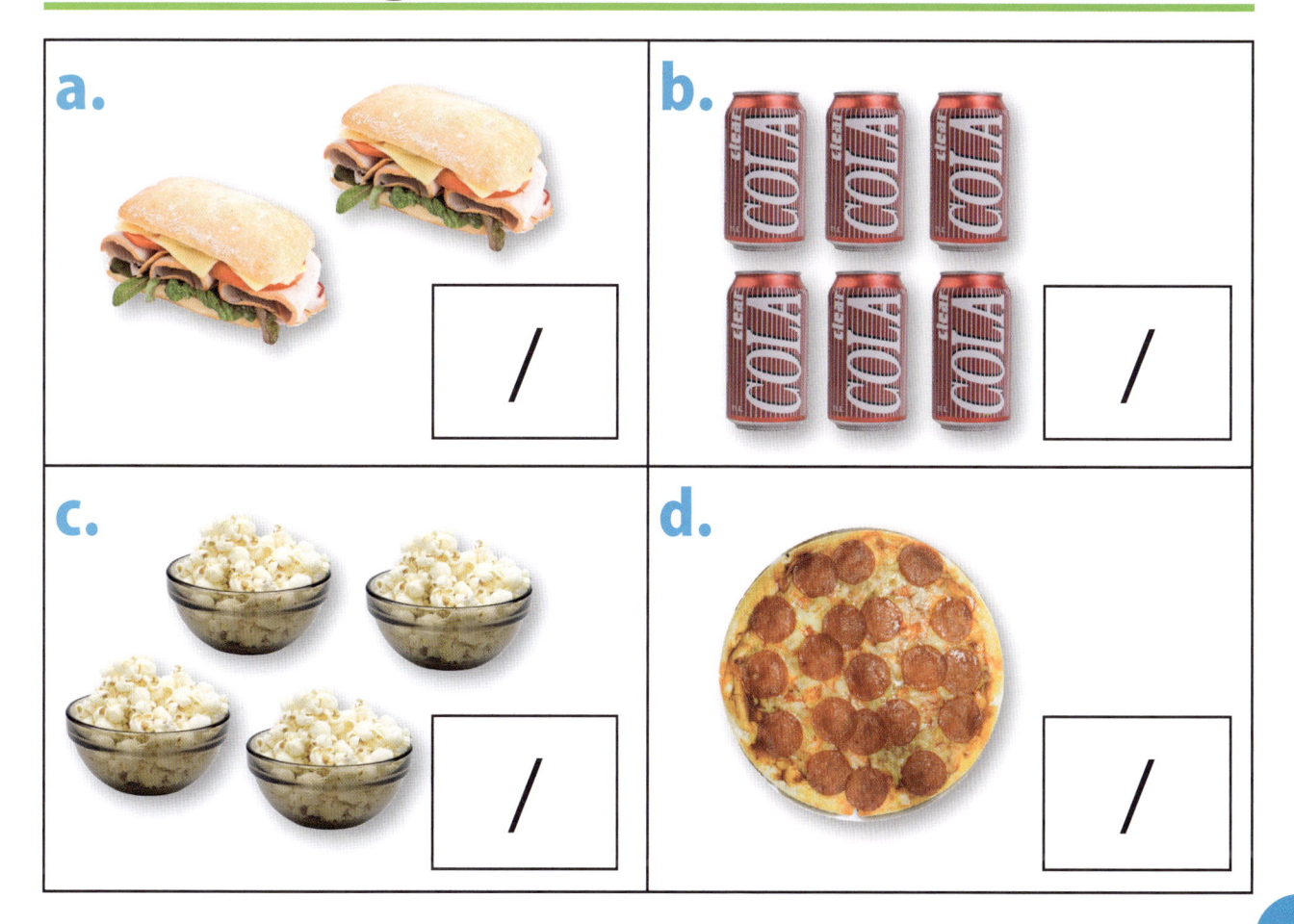

a. /

b. /

c. /

d. /

Chapter 6 • Fractions
159

Apple Pie

Date _____

Directions: Mrs. Perez baked an apple pie. Her family ate 1/2 of it for desert.
The family ate the other 1/2 for an evening snack.
How much of the pie was left? Wrire the fraction in the box.

/

Bonus: This coin is called a half-dollar. It is 1/2 of a 1.
◯ 1/2 of a 1.

Chapter 6 • Fractions

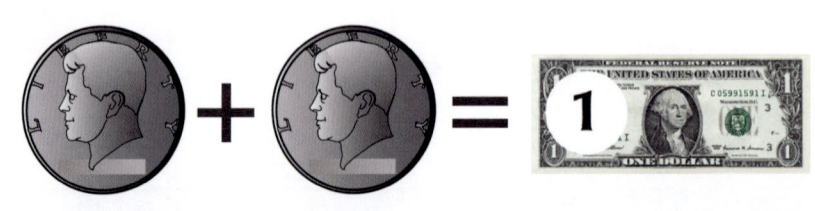

¼ of a Chocolate Bar

Date _____

Directions: Isabella wants to break her chocolate bar so that she can share it with 3 of her friends. Draw lines to show how to divide the chocolate bar evenly so that each person will get 1/4 of the candy bar. Shade 1/4 of the bar and write the fraction 1/4.

Hint: ▢▢ and ▢┊▢

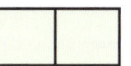

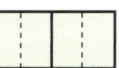

/

Bonus: A (quarter) is 1/4 of a $1. ◯ 1/4 of a $1.

(quarter) + (quarter) + (quarter) + (quarter) = $1

Chapter 6 • Fractions

Homecoming

Date _____

Directions: 4 cheerleaders will ride a float for homecoming. ◯ 1/4 of each set to show how much each cheerleader will get.

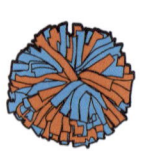

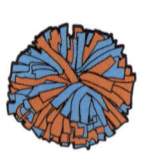

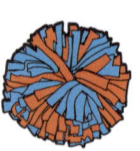

1

2

3

162 Chapter 6 • Fractions

Homemade Cookies

Date _____

Directions: Mrs. Perez made some cookies. She wants to put 1/2 of them in the cookie jar. She wants to put 1/4 on a plate for her family. The rest she wants to give to her neighbor.

1. Shade 1/2 of the cookies.
2. ◯ 1/4 of the cookies.
3. What fraction of the cookies will the neighbor get?

   _____

4. How many cookies will the neighbor get?

   _____

Chapter 6 • Fractions

More Pie

Date _____

Directions: Mrs. Perez is baking again. She wants to give 1/3 of the pie to her family for supper. 1/3 will be put into the refrigerator for snacks. She will give her neighbor 1/3. What fraction of the pie will Mrs. Perez keep for her family?
Shade the fraction and write it in the ☐.

/

Softball Practice

Date _____

Directions: Coach Peterson has 3 teams for softball practice. He has to divide the equipment evenly so that each team gets the same amount of equipment.
◯ the equipment so that it show 1/3 for each team.

a Sets of 5:

1/3 of 15 is _____

b Sets of 3:

1/3 of 9 is _____

c Sets of 6:

1/3 of 18 is _____

d Sets of 1:

1/3 of 3 is _____

Bonus: ◯ the 🪙s below to show 3/4 of a 💵.

Chapter 6 • Fractions

165

A Recipe

Date _____

Directions: Mrs. Perez is making a desert. Write the fraction that shows how much of each ingredient she needs.

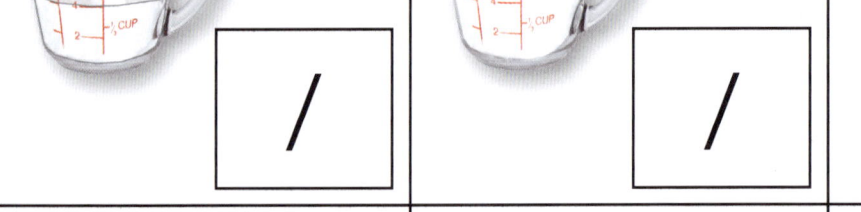

a. sugar

[/]

b. flour

[/]

c. chocolate syrup

[/]

d. butter

[/]

e. nuts

[/]

f. Bonus: eggs

[/]

Bonus: ◯ the 🪙 below to show 1/2 of a 💵 .

166

Chapter 6 • Fractions

Mini-Pizzas

Date _____

Directions: Mindy made mini-pizzas for a party. Shade the mini-pizzas and write the fraction to show the problem.

1 Mindy was hungry and ate 1/3 of a pizza before the party. Her sister came and ate another 1/3 of the same pizza. How much of the pizza did they eat altogether?

_____

2 Mindy's friend Irma took a pizza and ate 1/4 of it. After dancing she was hungry and ate another 2/4 of the pizza. How much did she eat altogether?

_____

3 Mindy gave 1/2 of a pizza to Melinda. How much of the pizza did Melinda not eat?

_____

Chapter 6 • Fractions

One-fifth

Date _____

Directions: Mrs. Perez made a cake for her family. She cut the cake into fifths. How much of the cake will each person get to eat? Shade each portion and write the fraction.

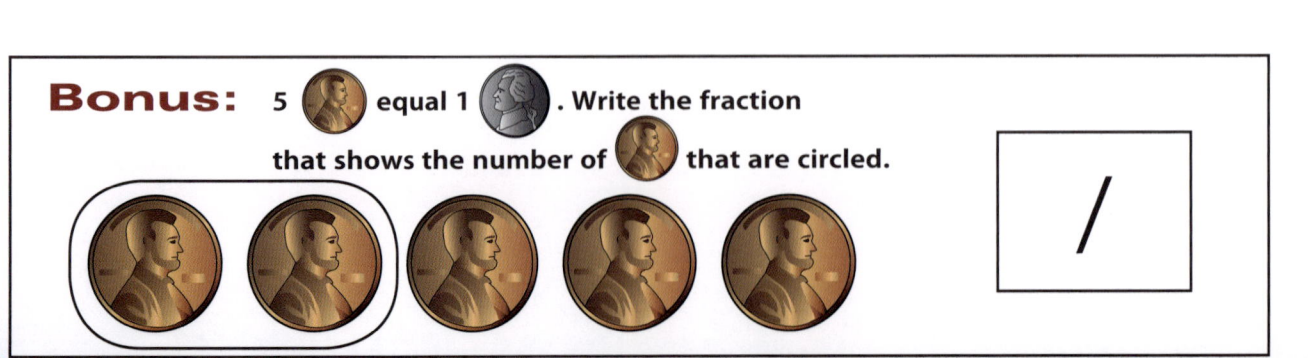

/

Bonus: 5 🪙 equal 1 🪙. Write the fraction that shows the number of 🪙 that are circled.

/

Chapter 6 • Fractions

Penny Candy Sales

Date _____

Directions: Henry and Ravi have 5¢, or a nickel's worth of pennies, to spend.
◯ the 🪙 and write the fraction to solve the problems below.
Hint: 5 🪙 = 🪙 and 🪙 = 1/5 of a 🪙.

1. Henry bought 1 🍭 and 3 🍬.

   How many 🪙 did Henry spend? _____

   What fraction of a nickel did he spend?

   🪙 🪙 🪙 🪙 🪙 = /

2. Ravi bought 2 🍬 and 1 🍭.

   How many 🪙 did Ravi spend? _____

   What fraction of a nickel did he spend?

   🪙 🪙 🪙 🪙 🪙 = /

Bonus: 5 🪙 = 🪙. ◯ 2/5 of a nickel.

🪙 🪙 🪙 🪙 🪙

Chapter 6 • Fractions

169

Garage Sale

Date _____

Directions: Henrietta and Rhonda each had 25¢ in [nickels]. They went to the nickel table at a garage sale. ◯ the [nickels] spent and write the fraction.

Hint: 5 [nickels] = [quarter] and [nickel] = 1/5 of a [quarter].

Everything costs a [nickel]!

1. Henrietta bought 1 [headband] and 1 [ring].

   How many [nickels] did Henrietta spend? _____

   What fraction of the 25¢ did she spend?

   [nickel] [nickel] [nickel] [nickel] [nickel] = ___/___

2. Rhonda bought 3 [magazines].

   How many [nickels] did Rhonda spend? _____

   What fraction of the 25¢ did she spend?

   [nickel] [nickel] [nickel] [nickel] [nickel] = ___/___

Bonus: Melissa spent 5 [nickels]. ◯ the nickels and tell how much she has left.

[nickel] [nickel] [nickel] [nickel] [nickel]

170

Chapter 6 • Fractions

Clothing Sale

Date _____

Directions: Miwok and Kasa each had $100.00 in [20] bills. They shopped for clothes on sale. ◯ the [20] spent and write the fraction.

Hint: 5 [20] = [100] and [20] = 1/5 of a [100].

---

1. Miwok bought a 🧥 and 👖.

   How many [20] did she spend? _____

   What fraction of the [100] did she spend?

   [20] [20] [20] [20] [20] = /

2. Kasa bought a 👜, 👠, and a 🧥.

   How many [20] did Kasa spend? _____

   What fraction of the [100] did she spend?

   [20] [20] [20] [20] [20] = /

Bonus: Count to [100] by 20.

[20] [20] [20] [20] [20]

Chapter 6 • Fractions

⅛ of a Cup

Date _____

Directions: Dominick wants to wash some clothes. He has to put in 1/8 of a cup of detergent. Shade and write the fraction that shows how much detergent Dominick needs.

= /

Bonus: You have 8 baseball cards and give 1/4 of them to a friend. How many do you have left?

_____

172

Chapter 6 • Fractions

Pieces of Melon

Date _____

Directions: Sonya cut a watermelon into 8 equal parts. She ate 3 pieces and Nicky ate 2 pieces. What fraction of the melon is left?

/

Bonus: Isabel has 3 Rock CDs and 1 Hip Hop CD. What fraction of her CDs are rock?

_____

Chapter 6 • Fractions

Hot Dog Stand

Date _____

Directions: Answer the questions below.

1 How many hot dogs are there? _____

2 What fraction is 1 hot dog?

3 What fraction of the hot dogs are plain?

4 What fraction of the hot dogs have catsup?

5 What fraction of the hot dogs have mustard?

6 What fraction of the hot dogs have sauerkraut?

Bonus: What fraction of the children have red hair? _____

Chapter 6 • Fractions

A Coin Collection

Date _____

Directions: Tyrone collects old 🪙 pennies. He has 14 old 🪙 pennies. His grandmother gave him 2 more. He now has 16 🪙 pennies. What fraction of the total amount of pennies did his grandmother give him?

_____

Bonus: What fraction of Tyrone's 🪙 are 8 🪙 pennies? _____

Chapter 6 • Fractions

1/10

Date _____

Directions: Isabel made a cake and cut it into 10 pieces. Shade 1/10th of the cake and write the fraction.

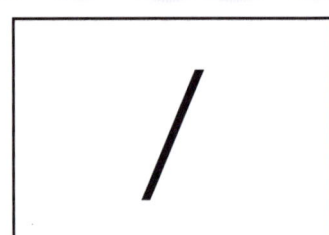

Bonus: 10 pennies = 1 dime. ◯ 1/10th of a dime.

Chapter 6 • Fractions

A Penny Collection

Date _____

Directions: Solve to find out what fraction of his penny collection Tyrone has left and write the fraction which shows that amount.

1 Tyrone has 10 new 🪙 pennies to put in his coin collection.

2 He gave 3 of the 🪙 pennies to his friend, who also collects coins.

3 ◯ the number of 🪙 pennies he has left.

4 Write the fraction that shows how many 🪙 pennies he has left.

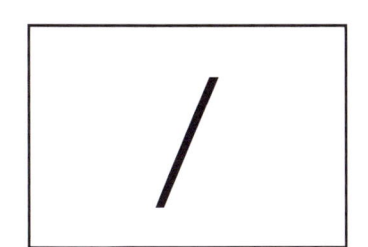

/

Chapter 6 • Fractions

A Breakfast Doughnut

Date _____

Directions: Solve to find out how much money Hilda has left after buying a doughnut. Write the fraction to show how much is left.

Hint: 10 dimes = $1 and dime = 1/10 of a $1

1 Hilda had 10 dimes that equal $1.

2 She buys a donut for $.60, or 6 dimes.

3 **X** the amount of dimes spent. ◯ the amount of dimes she has left.

4 Write the fraction to show how many dimes she has left.

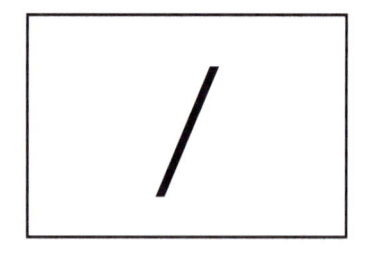

/

Bonus: 10 dimes = a $1 dollar. ◯ 1/10 of a $1 dollar.

Chapter 6 • Fractions

A New Cell Phone

Date _____

Directions: Solve to find out what fraction of her money Miwok has left after buying a cell phone, and write the fraction which shows that amount.

Hint: 10 [10] = [100] and [10] = 1/10 of a [100]

1 Miwok has 10 [10] ten-dollar bills that equal [100].

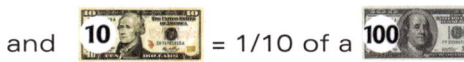

2 She buys a cell phone that costs $80.00, or 8 ten-dollar bills.

3 X out the amount of bills spent and ◯ the [10] ten-dollar bills she has left.

4 Write the fraction that shows how many bills she has left.

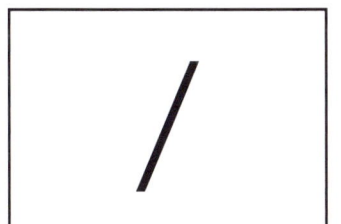

$$\boxed{/}$$

Bonus: 10 [10] ten-dollar bills = [100] a hundred-dollar bill.

◯ 1/10 of a [100] hundred-dollar bill.

Chapter 6 • Fractions

179

A Bagful of Candy

Date _____

Directions: Tracy bought some penny candy. What kind would he most likely pick? ◯ your answer. What fraction of the total is each kind of candy?

Write the fraction:

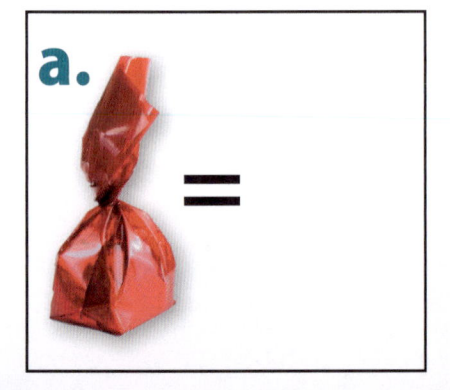

a. =

b. =

c. =

Chapter 6 • Fractions

Fraction Patterns

Date _____

Directions: Look at the pattern in each row. Shade the last one to show what comes next. Write the fraction of the shaded part under each figure.

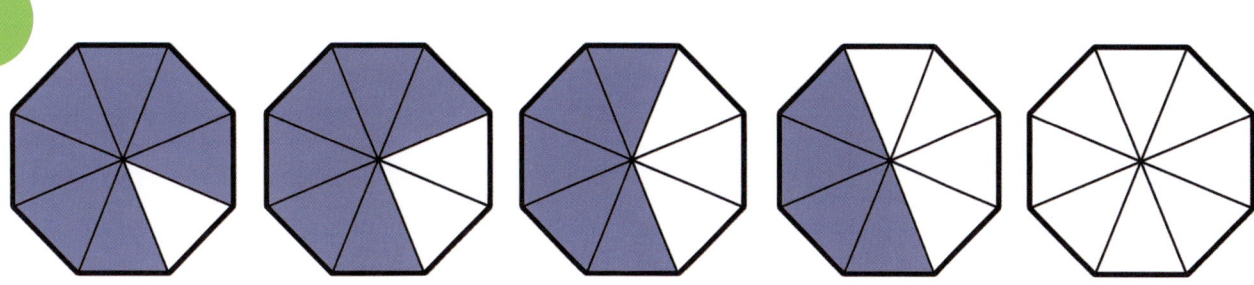

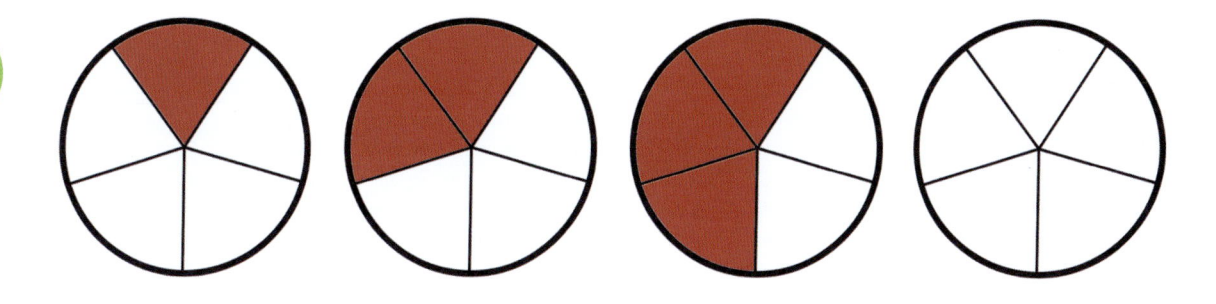

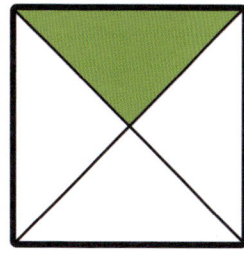

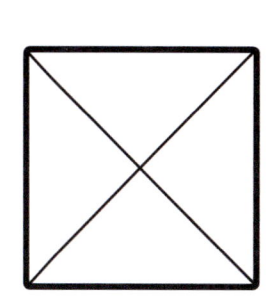

1

_____   _____   _____   _____   _____

2

_____   _____   _____   _____

3

_____   _____   _____   _____

Chapter 6 • Fractions

181

More Fractions 1

Date _____

Directions: Compare the size of learned fractions.

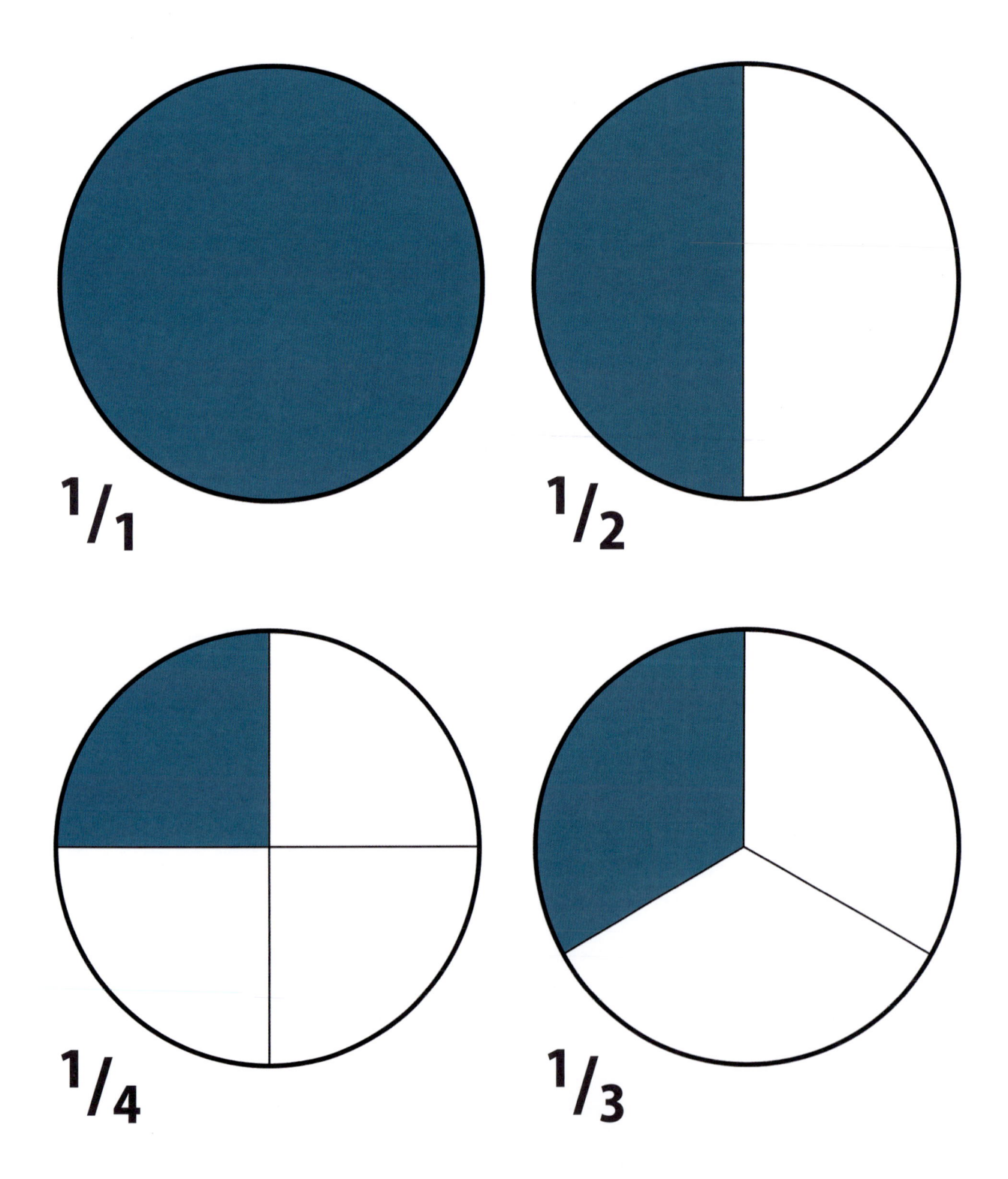

1/1

1/2

1/4

1/3

Chapter 6 • Fractions

More Fractions 2

Date _____

Directions: Compare the size of learned fractions.

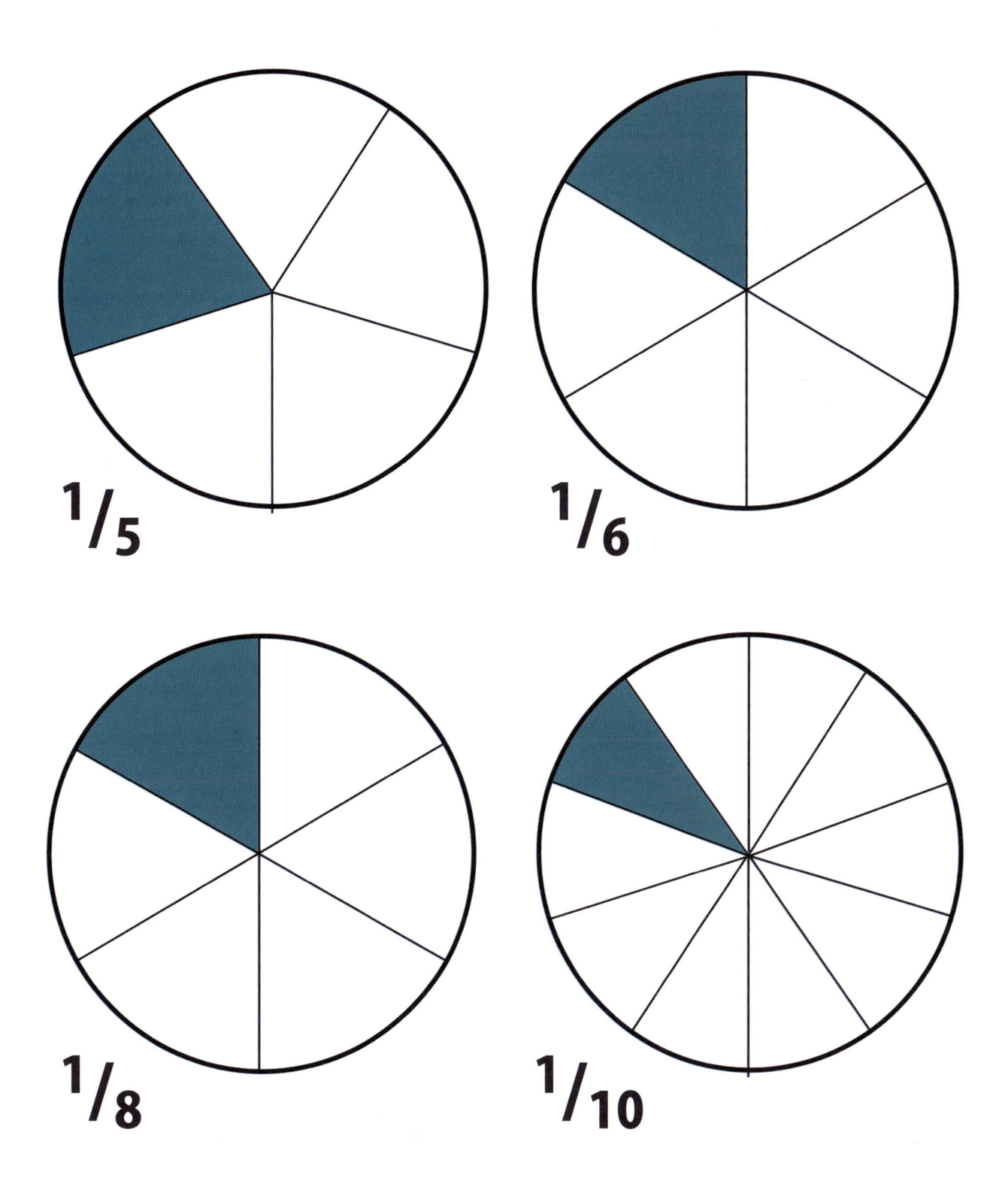

1/5

1/6

1/8

1/10

Chapter 6 • Fractions

183